I0002376

Federal Information Processing Standards Publication Stateless Hash-Based Digital Signature Standard

Contents

Federal Information Processing Standards Publication Stateless Hash-Based Digital Signature Standard

National Institute of Standards and Technology

Nimble Books LLC: The AI Lab for Book-Lovers

Fred Zimmerman, Editor

Humans and AI making books richer, more diverse, and more surprising

Publishing Information

- (c) 2024 Nimble Books LLC
- ISBN: 978-1-60888-335-6

Nimble Books LLC ~ NimbleBooks.com

Bibliographic Key Phrases

Stateless Hash-Based Digital Signature Standard; FIPS 205; NIST; Post-Quantum Cryptography; SPHINCS+; Quantum-Resistant Cryptography; Digital Signatures; Computer Security; Cryptography;

Publisher's Note

In an era marked by escalating cyber threats and the looming specter of quantum computing, safeguarding sensitive information is more critical than ever. This FIPS publication, NIST FIPS 205, stands as a critical bulwark in the face of these challenges. It introduces the **Stateless Hash-Based Digital Signature Algorithm (SLH-DSA)**, a cutting-edge digital signature scheme meticulously designed to withstand the threat posed by future quantum computers. This document delves into the intricate mathematical workings of SLH-DSA, providing a comprehensive guide to its key generation, signature generation, and signature verification processes. The publication further outlines 12 approved parameter sets, each offering distinct levels of security strength tailored to different application needs, and elucidating the precise instantiation of various hash functions and pseudorandom functions. This document also examines the intricate components that underpin the SLH-DSA scheme, including the Winternitz One-Time Signature Plus scheme (WOTS+), the eXtended Merkle Signature Scheme (XMSS), the Hypertree, and the Forest of Random Subsets (FORS). For those concerned with the implications of quantum computing on current cryptographic systems, this FIPS publication is a must-read. It offers a deep understanding of SLH-DSA's structure and function, empowering researchers, practitioners, and policymakers to navigate the complexities of post-quantum cryptography and implement this groundbreaking technology to secure our digital future.

This annotated edition illustrates the capabilities of the AI Lab for Book-Lovers to add context and ease-of-use to manuscripts. It includes publishing information; abstracts; viewpoints; learning aids; and references.

Truth in Publishing (Disclosures)

Context: This document, FIPS 205, outlines the Stateless Hash-Based Digital Signature Standard (SLH-DSA). It's an attempt to create a digital signature scheme resistant to quantum computers, a feat as ambitious as trying to convince a cat to take a bath.

Strengths:

- **Quantum-resistant:** It's supposed to be safe from those pesky quantum computers, though I wouldn't bet the farm on it just yet.
- **Detailed:** This thing is chock-full of technical details, which is great if you're a cryptography aficionado. If you're not, buckle up for a long ride.
- **Government-approved:** The official stamp of approval from NIST makes it seem all-important, even if it's about as exciting as a tax audit.

Weaknesses:

- **Dense:** The document is written in the most delightful, jargon-filled prose. Prepare to be bombarded with acronyms and technical terms that might leave you feeling like you're in a foreign language class.
- **Clunky:** The algorithms are presented in a way that's technically accurate but aesthetically challenged. Think of it as the equivalent of wearing a tie-dye shirt to a formal dinner.
- **Unreadable:** Let's be honest, unless you're deeply invested in the intricacies of cryptographic algorithms, this document is more likely to induce sleep than excitement.

Overall: FIPS 205 is a valiant attempt to address the security concerns of the digital world in the face of quantum computing. It's technically sound and well-intentioned, but it's as dry and engaging as a textbook on the history of paper clips.

Recommendation: Only read this if you're a cryptography enthusiast or have a morbid fascination with government regulations.

Analytic Table of Contents

Introduction: The Need for Post-Quantum Cryptography

- Overview of the vulnerability of current public-key cryptography to quantum attacks.
- Introduction of NIST's Post-Quantum Cryptography Standardization process and its goals.
- Explanation of the selection of SPHINCS+ (renamed SLH-DSA) as one of the first four post-quantum algorithms for standardization.

Chapter 1: SLH-DSA: A Stateless Hash-Based Signature Scheme

- Detailed breakdown of SLH-DSA's structure, highlighting its reliance on underlying hash-based signature schemes:
  - **Winternitz One-Time Signature Plus (WOTS+):** A one-time signature scheme with key generation, signature generation, and public key derivation functions.
  - **eXtended Merkle Signature Scheme (XMSS):** A multi-time signature scheme built upon WOTS+ and Merkle hash trees, with functions for generating XMSS signatures and public keys from signatures.
  - **Forest of Random Subsets (FORS):** A few-time signature scheme designed for signing message digests with gradually degrading security.
  - **Hypertree:** A hierarchical structure of XMSS trees that enables the efficient authentication of a large number of FORS keys.
- Explanation of the key generation process for SLH-DSA, emphasizing the role of random seeds and PRF keys.
- Description of the SLH-DSA signature generation process, including:
  - Randomized hashing of the message.
  - Selection of a FORS key based on the message digest.
  - Signing a portion of the message digest using the selected FORS key.
  - Generating a hypertree signature to authenticate the FORS key.
- Outline of the signature verification process, involving:
  - Computing a candidate FORS public key from the signature and message.
  - Verifying the hypertree signature on the FORS public key.
  - Validating the signature if the verification succeeds.

Chapter 2: Functions and Addressing

- Definition of the six core functions used in SLH-DSA, including hash functions and pseudorandom functions.
- Introduction of the 32-byte address (ADRS) concept and its role in domain separation and mitigation of attacks.
- Detailed description of the seven types of addresses used in SLH-DSA, emphasizing their structure and interpretation.
- Explanation of member functions for manipulating address components.
- Discussion of the representation of arrays, byte strings, and integers within the algorithm.
- Description of the base_2b function for converting messages into arrays of base-2b integers.

Chapter 3: Detailed Specifications of the Underlying Signature Schemes

- **Section 3.1: Winternitz One-Time Signature Plus (WOTS+):**
  - In-depth explanation of WOTS+ key generation, signature generation, and public key derivation functions.
  - Discussion of the chaining function used in WOTS+.

- **Section 3.2: eXtended Merkle Signature Scheme (XMSS):**
  - Detailed description of the XMSS tree generation process and the xmss_node function.
  - Explanation of the XMSS signature generation and public key derivation functions.
- **Section 3.3: Forest of Random Subsets (FORS):**
  - In-depth explanation of the FORS key generation process and the fors_skGen function.
  - Description of the Merkle tree generation process within FORS and the fors_node function.
  - Explanation of the FORS signature generation and public key derivation functions.
- **Section 3.4: Hypertree:**
  - Detailed description of the hypertree structure and its role in SLH-DSA.
  - Explanation of the ht_sign function for generating hypertree signatures.
  - Description of the ht_verify function for verifying hypertree signatures.

Chapter 4: Implementation Requirements and Considerations

- Detailed discussion of essential implementation requirements for secure SLH-DSA:
  - Randomness generation using approved random bit generators.
  - Secure destruction of sensitive data.
  - Key checks for validating public keys and ensuring private key possession.
  - Prohibition of floating-point arithmetic.
- Guidance on implementation considerations:
  - Avoiding support for component schemes (WOTS+, XMSS, FORS, and hypertree).
  - Mitigating side-channel and fault attacks.
  - Choosing appropriate hash functions and XOFs for pre-hashing.

Chapter 5: Approved Parameter Sets

- Presentation of the 12 approved parameter sets for SLH-DSA, including:
  - SHAKE-based parameter sets.
  - SHA2-based parameter sets for different security categories.
- Discussion of the security strength categories and their implications.
- Explanation of message-bound signatures and their potential limitations in SLH-DSA.
- Guidelines for choosing the appropriate parameter set for different applications.

Chapter 6: External Functions for Application Use

- Specification of the external functions for SLH-DSA key generation, signature generation, and signature verification.

- Discussion of the two variants of signature generation:
    - **Pure signing:** Signing the message directly, including domain separation information.
    - **Pre-hash signing:** Signing a hash of the message, including domain separation and hash function information.
- Guidance on choosing the appropriate signing variant for different use cases.
- Recommendations for secure use of external functions, including context management and randomness generation.

Appendix A: Changes From the SPHINCS+ Submission

- Summary of the modifications made to the SPHINCS+ specification in this standard.
- Explanation of the rationale behind these changes.

This Analytic Table of Contents provides a comprehensive outline of the FIPS 205 standard, providing insights into the technical details, security considerations, and implementation guidelines for the SLH-DSA algorithm.

Abstracts

TLDR (three words)

Post-quantum signature standard

ELI5

This is a super secret code that helps people know that a message really came from the person who sent it. Imagine you are writing a note to your friend, and you want to make sure that nobody changes it before your friend gets it. You could use a special pen that only you and your friend have, and you could sign your note with the pen. That special pen is like the secret code that this book tells you about. The secret code makes sure that nobody can change your message without you knowing about it. This secret code is a bit different because it's for computers, and it's super strong so that even bad guys with really powerful computers won't be able to break it.

Scientific-Style Abstract

This standard specifies the Stateless Hash-Based Digital Signature Algorithm (SLH-DSA), a digital signature scheme designed to be resistant to attacks from quantum computers. SLH-DSA is based on the SPHINCS+ algorithm, which was selected for standardization as part of the NIST Post-Quantum Cryptography Standardization Process. The standard defines the mathematical steps for key generation, signature generation, and signature verification, and it specifies several parameter sets for SLH-DSA that are approved for use. It also includes guidance for cryptographic modules that implement SLH-DSA, including requirements for randomness generation, destruction of sensitive data, and key checks. The standard addresses implementation considerations for SLH-DSA, such as the need to avoid floating-point arithmetic and the importance of protecting against side-channel and fault attacks. SLH-DSA is intended for use in applications that require digital signatures to ensure data integrity and origin authentication.

Learning Aids

Mnemonic (acronym)

SLH-DSA: **S**tateless **L**arge **H**ash-based **D**igital **S**ignature **A**lgorithm

Mnemonic (speakable)

Secure **L**arge **H**ash **D**igital **S**ignature **A**lgorithm

Mnemonic (singable)

To the tune of "Row, Row, Row Your Boat":

SLH-DSA, SLH-DSA, What do we use it for? To make signatures secure, And resistant to quantum war!

Excerpts

Most Important Passages.

1. Introduction (Page 1):

> This standard defines a method for digital signature generation that can be used for the protection of binary data (commonly called a message) and for the verification and validation of those digital signatures.

This passage clearly states the purpose and scope of the standard, emphasizing its focus on digital signature generation and validation.

2. Overview of the SLH-DSA Signature Scheme (Page 7):

> SLH-DSA is a stateless hash-based signature scheme that is constructed using other hash-based signature schemes as components: (1) a few-time signature scheme, forest of random subsets (FORS), and (2) a multi-time signature scheme, the eXtended Merkle Signature Scheme (XMSS). XMSS is constructed using the hash-based one-time signature scheme Winternitz One-Time Signature Plus (WOTS+) as a component.

This passage provides a concise overview of the SLH-DSA scheme's structure, outlining its dependence on other signature schemes like FORS and XMSS.

3. Additional Requirements (Page 9):

> Randomness generation. SLH-DSA key generation (Algorithm 21) requires the generation of three random -byte values: PK.seed, SK.seed, and SK.prf, where is 16, 24, or 32, depending on the parameter set. For each invocation of key generation, each of these values shall be a fresh (i.e., not previously used) random value generated using an approved random bit generator (RBG), as prescribed in SP 800-90A, SP 800-90B, and SP 800-90C [14, 15, 16]. Moreover, the RBG used shall have a security strength of at least 8 bits. See Table 2 for the value of for each parameter set.

This passage emphasizes the importance of randomness generation in key generation and mandates the use of approved RBGs with a specific security strength.

4. Implementation Considerations (Page 10):

> Do not support component use. As WOTS+, XMSS, FORS, and hypertree signature schemes are not approved for use as stand-alone signature schemes, cryptographic modules should not make interfaces to these components available to applications. SP 800-208 [12] specifies approved stateful hash-based signature schemes.

This passage warns implementers against providing direct access to the component schemes, emphasizing the standard's intent for SLH-DSA as a complete signature scheme.

5. SLH-DSA Key Generation (Page 33):

> The SLH-DSA public key contains two elements (see Figure 16). The first is an -byte public seed PK.seed, which is used in many hash function calls to provide domain separation between different SLH-DSA key pairs. The second value is the hypertree public key (i.e., the root of the top layer XMSS tree). PK.seed shall be generated using an approved random bit generator (see [14, 15, 16]), where the instantiation of the random bit generator supports at least 8 bits of security strength.

This passage explains the composition of the SLH-DSA public key and the requirements for generating the public seed, ensuring proper domain separation.

6. SLH-DSA Signature Verification (Page 36):

> As with signature generation, SLH-DSA signature verification (Algorithm 20) begins by computing a message digest (line 8) and then extracting (line 9), (lines 10 and 12), and (lines 11 and 13) from the digest. A candidate FORS public key is then computed (line 17), and the signature on the FORS key is verified (line 18). If this signature verification succeeds, then the correct FORS public key was computed, and the signature SIG on message is valid.

This passage outlines the steps involved in verifying an SLH-DSA signature, emphasizing the verification of the FORS public key as a key step in the process.

7. Parameter Sets (Page 43):

> This standard approves 12 parameter sets for use with SLH-DSA. A parameter set consists of parameters for WOTS+ (and), XMSS and the SLH-DSA hypertree (and), and FORS (and), as well as instantiations for the functions H , PRF, PRF , F, H, and T .

This passage introduces the concept of parameter sets within SLH-DSA, highlighting the various components and functions that are defined by each parameter set.

FIPS 205

Federal Information Processing Standards Publication

Stateless Hash-Based Digital Signature Standard

Category: Computer Security   **Subcategory: Cryptography**

Information Technology Laboratory
National Institute of Standards and Technology
Gaithersburg, MD 20899-8900

This publication is available free of charge from:
https://doi.org/10.6028/NIST.FIPS.205

Published: August 13, 2024

U.S. Department of Commerce
Gina M. Raimondo, Secretary

National Institute of Standards and Technology
Laurie E. Locascio, NIST Director and Under Secretary of Commerce for Standards and Technology

Foreword

The Federal Information Processing Standards Publication (FIPS) series of the National Institute of Standards and Technology (NIST) is the official series of publications relating to standards and guidelines developed under 15 U.S.C. 278g-3, and issued by the Secretary of Commerce under 40 U.S.C. 11331.

Comments concerning this Federal Information Processing Standard publication are welcomed and should be submitted using the contact information in the "Inquiries and comments" clause of the announcement section.

Kevin M. Stine, Director
Information Technology Laboratory

Abstract

This standard specifies the stateless hash-based digital signature algorithm (SLH-DSA). Digital signatures are used to detect unauthorized modifications to data and to authenticate the identity of the signatory. In addition, the recipient of signed data can use a digital signature as evidence in demonstrating to a third party that the signature was, in fact, generated by the claimed signatory. This is known as non-repudiation since the signatory cannot easily repudiate the signature at a later time. SLH-DSA is based on SPHINCS$^+$, which was selected for standardization as part of the NIST Post-Quantum Cryptography Standardization process.

Keywords: computer security; cryptography; digital signatures; Federal Information Processing Standards; hash-based signatures; post-quantum; public-key cryptography.

Federal Information Processing Standards Publication 205

Published: August 13, 2024
Effective: August 13, 2024

Announcing the

Stateless Hash-Based Digital Signature Standard

Federal Information Processing Standards (FIPS) publications are developed by the National Institute of Standards and Technology (NIST) under 15 U.S.C. 278g-3 and issued by the Secretary of Commerce under 40 U.S.C. 11331.

1. **Name of Standard.** Stateless Hash-Based Digital Signature Standard (FIPS 205).

2. **Category of Standard.** Computer Security. **Subcategory.** Cryptography.

3. **Explanation.** This standard specifies a stateless hash-based digital signature scheme (SLH-DSA) for applications that require a digital signature rather than a written signature. Additional digital signature schemes are specified and approved in other NIST Special Publications and FIPS publications (e.g., FIPS 186-5 [1]). A digital signature is represented in a computer as a string of bits and computed using a set of rules and parameters that allow the identity of the signatory and the integrity of the data to be verified. Digital signatures may be generated on both stored and transmitted data.

    Signature generation uses a private key to generate a digital signature. Signature verification uses a public key that corresponds to but is not the same as the private key. Each signatory possesses a private and public key pair. Public keys may be known by the public, but private keys must be kept secret. Anyone can verify the signature by employing the signatory's public key. Only the user who possesses the private key can perform signature generation.

    The digital signature is provided to the intended verifier along with the signed data. The verifying entity verifies the signature by using the claimed signatory's public key. Similar procedures may be used to generate and verify signatures for both stored and transmitted data.

    This standard specifies several parameter sets for SLH-DSA that are **approved** for use. Additional parameter sets may be specified and approved in future NIST Special Publications.

4. **Approving Authority.** Secretary of Commerce.

5. **Maintenance Agency.** Department of Commerce, National Institute of Standards and Technology, Information Technology Laboratory (ITL).

6. **Applicability.** This standard is applicable to all federal departments and agencies for the protection of sensitive unclassified information that is not subject to section 2315 of Title 10, United States Code, or section 3502 (2) of Title 44, United States Code. Either this standard, FIPS 204, FIPS 186-5, or NIST Special Publication 800-208 **shall** be used in designing and implementing public-key-based signature systems that federal departments and agencies operate or that are operated for them under contract. In the future, additional digital signature

i

schemes may be specified and approved in FIPS publications or NIST Special Publications.

The adoption and use of this standard are available to private and commercial organizations.

7. **Applications.** A digital signature algorithm allows an entity to authenticate the integrity of signed data and the identity of the signatory. The recipient of a signed message can use a digital signature as evidence in demonstrating to a third party that the signature was, in fact, generated by the claimed signatory. This is known as non-repudiation since the signatory cannot easily repudiate the signature at a later time. A digital signature algorithm is intended for use in electronic mail, electronic funds transfer, electronic data interchange, software distribution, data storage, and other applications that require data integrity assurance and data origin authentication.

8. **Implementations.** A digital signature algorithm may be implemented in software, firmware, hardware, or any combination thereof. NIST will develop a validation program to test implementations for conformance to the algorithms in this standard. For every computational procedure that is specified in this standard, a conforming implementation may replace the given set of steps with any mathematically equivalent process. In other words, different procedures that produce the correct output for every input are permitted. Information about validation programs is available at https://csrc.nist.gov/projects/cmvp. Examples for digital signature algorithms are available at https://csrc.nist.gov/projects/cryptographic-standards-and-guidelines/example-values.

Agencies are advised that digital signature key pairs **shall not** be used for other purposes.

9. **Other Approved Security Functions.** Digital signature implementations that comply with this standard **shall** employ cryptographic algorithms that have been **approved** for protecting Federal Government-sensitive information. **Approved** cryptographic algorithms and techniques include those that are either:

    a. Specified in a Federal Information Processing Standard (FIPS) publication,

    b. Adopted in a FIPS or NIST recommendation, or

    c. Specified in the list of **approved** security functions in SP 800-140C.

10. **Export Control.** Certain cryptographic devices and technical data regarding them are subject to federal export controls. Exports of cryptographic modules that implement this standard and technical data regarding them must comply with these federal regulations and be licensed by the Bureau of Industry and Security of the U.S. Department of Commerce. Information about export regulations is available at https://www.bis.doc.gov.

11. **Patents.** The algorithm in this standard may be covered by U.S. or foreign patents.

12. **Implementation Schedule.** This standard becomes effective immediately upon final publication.

13. **Specifications.** Federal Information Processing Standard (FIPS) 205, Stateless Hash-Based Digital Signature Standard (affixed).

14. **Qualifications.** The security of a digital signature system depends on the secrecy of the signatory's private keys. Signatories **shall**, therefore, guard against the disclosure of their private keys. While it is the intent of this standard to specify general security requirements for generating digital signatures, conformance to this standard does not ensure that a particular implementation is secure. It is the responsibility of an implementer to ensure that any module that implements a digital signature capability is designed and built in a secure manner.

Similarly, the use of a product containing an implementation that conforms to this standard does not guarantee the security of the overall system in which the product is used. The responsible authority in each agency or department **shall** ensure that an overall implementation provides an acceptable level of security.

Since a standard of this nature must be flexible enough to adapt to advancements and innovations in science and technology, this standard will be reviewed every five years in order to assess its adequacy.

15. **Waiver Procedure.** The Federal Information Security Management Act (FISMA) does not allow for waivers to Federal Information Processing Standards (FIPS) that are made mandatory by the Secretary of Commerce.

16. **Where to Obtain Copies of the Standard.** This publication is available by accessing https://csrc.nist.gov/publications. Other computer security publications are available at the same website.

17. **How to Cite This Publication.** NIST has assigned **NIST FIPS 205** as the publication identifier for this FIPS, per the NIST Technical Series Publication Identifier Syntax. NIST recommends that it be cited as follows:

> National Institute of Standards and Technology (2024) Stateless Hash-Based Digital Signature Standard. (Department of Commerce, Washington, D.C.), Federal Information Processing Standards Publication (FIPS) NIST FIPS 205. https://doi.org/10.6028/NIST.FIPS.205

18. **Inquiries and Comments.** Inquiries and comments about this FIPS may be submitted to fips-205-comments@nist.gov.

Federal Information Processing Standards Publication 205

Specification for the
Stateless Hash-Based Digital Signature Standard

Table of Contents

List of Tables

List of Figures

List of Algorithms

1. Introduction

1.1 Purpose and Scope

This standard defines a method for digital signature generation that can be used for the protection of binary data (commonly called a message) and for the verification and validation of those digital signatures.[1] The security of the stateless hash-based digital signature algorithm (SLH-DSA) relies on the presumed difficulty of finding preimages for hash functions as well as several related properties of the same hash functions. Unlike the algorithms specified in FIPS 186-5 [1], SLH-DSA is designed to provide resistance against attacks from a large-scale quantum computer.

This standard specifies the mathematical steps that need to be performed for key generation, signature generation, and signature verification. Additional assurances are required for digital signatures to be valid (e.g., the assurance of identity and private key possession). SP 800-89, *Recommendation for Obtaining Assurances for Digital Signature Applications* [3], specifies the required assurances and the methods for obtaining these assurances.

1.2 Context

Over the past several years, there has been steady progress toward building quantum computers. The security of many commonly used public-key cryptosystems will be at risk if large-scale quantum computers are ever realized. This would include key-establishment schemes and digital signatures that are based on integer factorization and discrete logarithms (both over finite fields and elliptic curves). As a result, in 2016, NIST initiated a public Post-Quantum Cryptography (PQC) Standardization process to select quantum-resistant public-key cryptographic algorithms for standardization. A total of 82 candidate algorithms were submitted to NIST for consideration.

After three rounds of evaluation and analysis, NIST selected the first four algorithms for standardization. These algorithms are intended to protect sensitive U.S. Government information well into the foreseeable future, including after the advent of cryptographically relevant quantum computers. This standard includes the specification for one of the algorithms selected: SPHINCS$^+$, a stateless hash-based digital signature scheme. This standard contains several minor modifications compared to Version 3 [4], which was submitted at the beginning of round three of the NIST PQC Standardization process. The changes are described in Appendix A. Throughout this standard, SPHINCS$^+$ will be referred to as *SLH-DSA* for stateless hash-based digital signature algorithm.

---

[1]NIST Special Publication (SP) 800-175B [2], *Guideline for Using Cryptographic Standards in the Federal Government: Cryptographic Mechanisms*, includes a general discussion of digital signatures.

2. Glossary of Terms, Acronyms, and Symbols

2.1 Terms and Definitions

approved	FIPS-approved and/or NIST-recommended. An algorithm or technique that is either 1) specified in a FIPS or NIST recommendation, 2) adopted in a FIPS or NIST recommendation, or 3) specified in a list of NIST-**approved** security functions. [1]
big-endian	The property of a byte string having its bytes positioned in order of decreasing significance. In particular, the leftmost (first) byte is the most significant, and the rightmost (last) byte is the least significant. The term "big-endian" may also be applied in the same manner to bit strings. [5, adapted]
byte string	An array of integers in which each integer is in the set $\{0, \dots, 255\}$.
claimed signatory	From the verifier's perspective, the claimed signatory is the entity that purportedly generated a digital signature. [1]
destroy	An action applied to a key or a piece of secret data. After a key or a piece of secret data is destroyed, no information about its value can be recovered. [1]
digital signature	The result of a cryptographic transformation of data that, when properly implemented, provides a mechanism for verifying origin authentication, data integrity, and signatory non-repudiation. [1]
entity	An individual (person), organization, device, or process. Used interchangeably with *party*. [1]
equivalent process	Two processes are equivalent if the same output is produced when the same values are input to each process (either as input parameters, as values made available during the process, or both). [1]
extendable-output function	A function on bit strings in which the output can be extended to any desired length. **Approved** XOFs (such as those specified in FIPS 202 [6]) are designed to satisfy the following properties as long as the specified output length is sufficiently long to prevent trivial attacks: 1. (One-way) It is computationally infeasible to find any input that maps to any new pre-specified output. 2. (Collision-resistant) It is computationally infeasible to find any two distinct inputs that map to the same output. [7, adapted]
fresh random value	A previously unused output of a random bit generator.
hash function	A function on bit strings in which the length of the output is fixed. **Approved** hash functions (such as those specified in FIPS 180 [8] and FIPS 202 [6]) are designed to satisfy the following properties:

1. (One-way) It is computationally infeasible to find any input that maps to any new pre-specified output

2. (Collision-resistant) It is computationally infeasible to find any two distinct inputs that map to the same output. [1]

hash value	See *message digest*.
key	A parameter used in conjunction with a cryptographic algorithm that determines its operation. Examples applicable to this standard include:

1. The computation of a digital signature from data, and

2. The verification of a digital signature. [1]

key pair	A public key and its corresponding private key. [1]
message	The data that is signed. Also known as *signed data* during the signature verification and validation process. [1]
message digest	The result of applying a hash function to a message. Also known as a *hash value*. [1]
non-repudiation	A service that is used to provide assurance of the integrity and origin of data in such a way that the integrity and origin can be verified and validated by a third party as having originated from a specific entity in possession of the private key (i.e., the signatory). [1]
owner	A key pair owner is the entity authorized to use the private key of a key pair. [1]
party	An individual (person), organization, device, or process. Used interchangeably with *entity*. [1]
private key	A cryptographic key that is used with an asymmetric (public-key) cryptographic algorithm. The private key is uniquely associated with the owner and is not made public. The private key is used to compute a digital signature that may be verified using the corresponding public key. [1]
pseudorandom	A process or data produced by a process is said to be pseudorandom when the outcome is deterministic yet also effectively random as long as the internal action of the process is hidden from observation. For cryptographic purposes, "effectively random" means "computationally indistinguishable from random within the limits of the intended security strength." [1]
public key	A cryptographic key that is used with an asymmetric (public-key) cryptographic algorithm and is associated with a private key. The public key is associated with an owner and may be made public. In the case of digital signatures, the public key is used to verify a digital signature that was generated using the corresponding private key. [1]

security category	A number associated with the security strength of a post-quantum cryptographic algorithm, as specified by NIST (see [9, Sect. 5.6]).
security strength	A number associated with the amount of work (i.e., the number of operations) that is required to break a cryptographic algorithm or system. [1]
shall	Used to indicate a requirement of this standard. [1]
should	Used to indicate a strong recommendation but not a requirement of this standard. Ignoring the recommendation could result in undesirable results. [1]
signatory	The entity that generates a digital signature on data using a private key. [1]
signature generation	The process of using a digital signature algorithm and a private key to generate a digital signature on data. [1]
signature validation	The (mathematical) verification of the digital signature and obtaining the appropriate assurances (e.g., public-key validity, private-key possession, etc.). [1]
signature verification	The process of using a digital signature algorithm and a public key to verify a digital signature on data. [1]
signed data	The data or message upon which a digital signature has been computed. Also see *message*. [1]
verifier	The entity that verifies the authenticity of a digital signature using the public key. [1]

2.2 Acronyms

ADRS	Address
ADRS^c	Compressed Address
AES	Advanced Encryption Standard
DER	Distinguished Encoding Rules
FIPS	Federal Information Processing Standard
FORS	Forest of Random Subsets
ITL	Information Technology Laboratory
MGF	Mask Generation Function
NIST	National Institute of Standards and Technology
OID	Object Identifier
PQC	Post-Quantum Cryptography
PRF	Pseudorandom Function

SHA	Secure Hash Algorithm
SHAKE	Secure Hash Algorithm KECCAK
SP	Special Publication
RFC	Request for Comments
WOTS$^+$	Winternitz One-Time Signature Plus
XMSS	eXtended Merkle Signature Scheme
XOF	eXtendable-Output Function

2.3 Mathematical Symbols

$X \parallel Y$ The concatenation of two arrays X and Y. If X is an array of length ℓ_x, and Y is an array of length ℓ_y, then $Z = X \parallel Y$ is an array of length $\ell_x + \ell_y$ such that

$$Z[i] = \begin{cases} X[i] & \text{if } 0 \leq i < \ell_x \\ Y[i - \ell_x] & \text{if } \ell_x \leq i < \ell_x + \ell_y. \end{cases}$$

$X[i : j]$ A subarray of X. If X is an array of length ℓ_x, $0 \leq i < j \leq \ell_x$, and $Y = X[i : j]$, then Y is an array of length $j - i$ such that $Y[k] = X[i + k]$ for $0 \leq k < j - i$.

$\mathsf{Trunc}_\ell(X)$ A truncation function that outputs the leftmost ℓ bytes of the input byte string X. If $Y = \mathsf{Trunc}_\ell(X)$, then Y is a byte string (array) of length ℓ such that $Y[i] = X[i]$ for $0 \leq i < \ell$ (i.e., $Y = X[0 : \ell]$).

$|X|$ The length (in bytes) of byte string X.

$\lceil a \rceil$ The ceiling of a; the smallest integer that is greater than or equal to a. For example, $\lceil 5 \rceil = 5$, $\lceil 5.3 \rceil = 6$, and $\lceil -2.1 \rceil = -2$. [1]

$\lfloor a \rfloor$ The floor of a; the largest integer that is less than or equal to a. For example, $\lfloor 5 \rfloor = 5$, $\lfloor 5.3 \rfloor = 5$, and $\lfloor -2.1 \rfloor = -3$. [1]

$a \bmod n$ The unique remainder r, $0 \leq r \leq (n - 1)$, when integer a is divided by the positive integer n. For example, $23 \bmod 7 = 2$. [1]

$a \cdot b$ The product of a and b. For example, $3 \cdot 5 = 15$.

a^b a raised to the power b. For example, $2^5 = 32$.

$\log_2 x$ The base 2 logarithm of x. For example, $\log_2(16) = 4$.

0b   The prefix to a number that is represented in binary.

0x   The prefix to a number that is represented in hexadecimal. [1, adapted]

$a \gg b$ The logical right shift of a by b positions (i.e., $a \gg b = \lfloor a/2^b \rfloor$). For example, $\mathsf{0x73} \gg 4 = 7$. [10, adapted]

$a \ll b$	The logical left shift of a by b positions (i.e., $a \ll b = a \cdot 2^b$). For example, $0x73 \ll 4 = 0x730$. [10, adapted]
$a \oplus b$	The bitwise exclusive-or of a and b. For example, $115 \oplus 1 = 114$ ($115 \oplus 1 = 0b01110011 \oplus 0b00000001 = 0b01110010 = 114$).
$s \leftarrow x$	In pseudocode, this notation means that the variable s is set to the value of the expression x.
$s \xleftarrow{\$} \mathbb{B}^n$	In pseudocode, this notation means that the variable s is set to a byte string of length n chosen at random. A fresh random value is generated for each time this step is performed.
$\perp$	A symbol indicating failure or the lack of output from an algorithm.

3. Overview of the SLH-DSA Signature Scheme

SLH-DSA is a stateless hash-based signature scheme that is constructed using other hash-based signature schemes as components: (1) a few-time signature scheme, forest of random subsets (FORS), and (2) a multi-time signature scheme, the eXtended Merkle Signature Scheme (XMSS). XMSS is constructed using the hash-based one-time signature scheme Winternitz One-Time Signature Plus (WOTS$^+$) as a component.[2]

Conceptually, an SLH-DSA key pair consists of a very large set of FORS key pairs.[3] The few-time signature scheme FORS allows each key pair to safely sign a small number of messages. An SLH-DSA signature is created by computing a randomized hash of the message, using part of the resulting message digest to pseudorandomly select a FORS key, and signing the remaining part of the message digest with that key. An SLH-DSA signature consists of the FORS signature and the information that authenticates the FORS public key. The authentication information is created using XMSS signatures.

XMSS is a multi-time signature scheme that is created using a combination of WOTS$^+$ one-time signatures and Merkle hash trees [13]. An XMSS key consists of $2^{h'}$ WOTS$^+$ keys and can sign $2^{h'}$ messages. The WOTS$^+$ public keys are formed into a Merkle hash tree, and the root of the tree is the XMSS public key. (The Merkle hash tree formed from the WOTS$^+$ keys is also referred to as an XMSS tree.) An XMSS signature consists of a WOTS$^+$ signature and an authentication path within the Merkle hash tree for the WOTS$^+$ public key. In Figure 1, triangles represent XMSS trees, squares represent the WOTS$^+$ public keys, and circles represent the interior nodes of the hash tree. Within an XMSS tree, the square and circles that are filled in represent the authentication path for the WOTS$^+$ public key needed to verify the signature.

The authentication information for a FORS public key is a hypertree signature. A hypertree is a tree of XMSS trees, as depicted in Figure 1. The tree consists of d layers[4] in which the top layer (layer $d-1$) consists of a single XMSS tree, the next layer down (layer $d-2$) consists of $2^{h'}$ XMSS trees, and the lowest layer (layer 0) consists of $2^{(d-1)h'}$ XMSS trees. The public key of each XMSS key at layers 0 through $d-2$ is signed by an XMSS key at the next higher layer. The XMSS keys at layer 0 collectively have $2^{dh'} = 2^h$ WOTS$^+$ keys, which are used to sign the 2^h FORS public keys in the SLH-DSA key pair. The sequence of d XMSS signatures needed to authenticate a FORS public key when starting with the public key of the XMSS key at layer $d-1$ is a hypertree signature. An SLH-DSA signature consists of a FORS signature along with a hypertree signature.

An SLH-DSA public key (Figure 16) contains two n-byte components: (1) $\mathbf{PK}.$root, which is the public key of the XMSS key at layer $d-1$, and (2) $\mathbf{PK}.$seed, which is used to provide domain separation between different SLH-DSA key pairs. An SLH-DSA private key (Figure 15) consists of an n-byte seed $\mathbf{SK}.$seed that is used to pseudorandomly generate all of the secret values for the WOTS$^+$ and FORS keys and an n-byte key $\mathbf{SK}.$prf that is used in the generation of the randomized hash of the message. An SLH-DSA private key also includes copies of $\mathbf{PK}.$root and $\mathbf{PK}.$seed, as these values are needed during both signature generation and signature verification.

---

[2]The WOTS$^+$ and XMSS schemes that are used as components of SLH-DSA are not the same as the WOTS$^+$ and XMSS schemes in RFC 8391 [11] and SP 800-208 [12].
[3]For the parameter sets in this standard, an SLH-DSA key pair contains 2^{63}, 2^{64}, 2^{66}, or 2^{68} FORS keys, which are pseudorandomly generated from a single seed.
[4]For the parameter sets in this standard, d is 7, 8, 17, or 22.

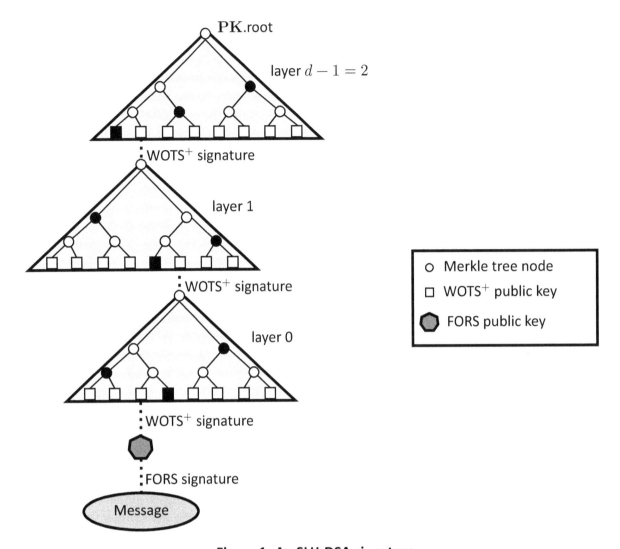

Figure 1. An SLH-DSA signature

The WOTS$^+$ one-time signature scheme is specified in Section 5, and the XMSS multi-time signature scheme is specified in Section 6. Section 7 specifies the generation and verification of hypertree signatures. The FORS few-time signature scheme is specified in Section 8. Finally, Section 9 specifies the SLH-DSA key generation, signature, and verification functions. As the WOTS$^+$, XMSS, hypertree, and FORS schemes described in this standard are not intended for use as stand-alone signature schemes, only the components of the schemes necessary to implement SLH-DSA are described. In particular, these sections do not include functions for key pair generation, and a signature verification function is only specified for hypertree signatures.

When used in this standard, WOTS$^+$, XMSS, and FORS signatures are implicitly verified using functions to generate public keys from messages and signatures (see Sections 5.3, 6.3, and 8.4). When verifying an SLH-DSA signature, the randomized hash of the message and the FORS signature are used to compute a candidate FORS public key. The candidate FORS public key and the WOTS$^+$ signature from the layer 0 XMSS key are used to compute a candidate WOTS$^+$ public key, which is then used in conjunction with the corresponding authentication path to compute a candidate XMSS public key. The candidate layer 0 XMSS public key is used along with the layer 1

XMSS signature to compute a candidate layer 1 XMSS public key. This process is repeated until a candidate layer $d - 1$ public key has been computed. SLH-DSA signature verification succeeds if the computed candidate layer $d - 1$ XMSS public key is the same as the SLH-DSA public key root $\mathbf{PK}$.root.

3.1   Additional Requirements

This section specifies requirements for cryptographic modules that implement SLH-DSA. Section 3.2 discusses issues that implementers of cryptographic modules should take into consideration but that are not requirements. SP 800-89, *Recommendation for Obtaining Assurances for Digital Signature Applications* [3], specifies requirements that apply to the use of digital signature schemes.

Randomness generation. SLH-DSA key generation (Algorithm 21) requires the generation of three random n-byte values: $\mathbf{PK}$.seed, $\mathbf{SK}$.seed, and $\mathbf{SK}$.prf, where n is 16, 24, or 32, depending on the parameter set. For each invocation of key generation, each of these values **shall** be a fresh (i.e., not previously used) random value generated using an **approved** random bit generator (RBG), as prescribed in SP 800-90A, SP 800-90B, and SP 800-90C [14, 15, 16]. Moreover, the RBG used **shall** have a security strength of at least $8n$ bits. See Table 2 for the value of n for each parameter set.

Destruction of sensitive data. Data used internally by key generation and signing algorithms in intermediate computation steps could be used by an adversary to gain information about the private key and thereby compromise security. The data used internally by verification algorithms is similarly sensitive for some applications, including the verification of signatures that are used as bearer tokens (i.e., authentication secrets) or signatures on plaintext messages that are intended to be confidential. Intermediate values of the verification algorithm may reveal information about its inputs (i.e., the message, signature, and public key), and in some applications, security or privacy requires one or more of these inputs to be confidential. Therefore, implementations of SLH-DSA **shall** ensure that any local copies of the inputs and any potentially sensitive intermediate data are destroyed as soon as they are no longer needed.

Key checks. SP 800-89 imposes requirements for the assurance of public-key validity and private-key possession. In the case of SLH-DSA, where public-key validation is required, implementations **shall** verify that the public key is $2n$ bytes in length. When the assurance of private key possession is obtained via regeneration, the owner of the private key **shall** check that the private key is $4n$ bytes in length and **shall** use $\mathbf{SK}$.seed and $\mathbf{PK}$.seed to recompute $\mathbf{PK}$.root and compare the newly generated value with the value in the private key currently held.

Floating-point arithmetic. Implementations of SLH-DSA **shall not** use floating-point arithmetic, as rounding errors in floating point operations may lead to incorrect results in some cases. In all pseudocode in this standard in which division is performed (e.g., x/y) and y may not divide x, either $\lfloor x/y \rfloor$ or $\lceil x/y \rceil$ is used. Both of these can be computed without floating-point arithmetic, as ordinary integer division x/y computes $\lfloor x/y \rfloor$, and $\lceil x/y \rceil = \lfloor (x+y-1)/y \rfloor$ for non-negative integers x and positive integers y.

While the value of len_2 (see Equation 5.3) may be computed without using floating-point arithmetic (see Algorithm 1), it is recommended that this value be precomputed. For all parameter sets in this standard, len_2 is 3.

Algorithm 1 gen_len$_2(n, lg_w)$

Computes len_2 (Equation 5.3).

Input: Security parameter n, bits per hash chain lg_w.
Output: len_2.

1:   $w \leftarrow 2^{lg_w}$ $\triangleright$ Equation 5.1
2:   $len_1 \leftarrow \left\lfloor \frac{8 \cdot n + lg_w - 1}{lg_w} \right\rfloor$ $\triangleright$ Equation 5.2
3:   $max\_checksum = len_1 \cdot (w - 1)$ $\triangleright$ maximum possible checksum value
4:   $len_2 \leftarrow 1$ $\triangleright$ maximum value that may be signed using
5:   $capacity \leftarrow w$ $\triangleright len_2$ hash chains is $w^{len_2} - 1 = capacity - 1$
6:   **while** $capacity \leq max\_checksum$ **do**
7:       $len_2 \leftarrow len_2 + 1$
8:       $capacity \leftarrow capacity \cdot w$
9:   **end while**
10:   **return** len_2

3.2 Implementation Considerations

This section discusses some implementation considerations for SLH-DSA.

Do not support component use. As WOTS$^+$, XMSS, FORS, and hypertree signature schemes are not approved for use as stand-alone signature schemes, cryptographic modules **should not** make interfaces to these components available to applications. SP 800-208 [12] specifies **approved** stateful hash-based signature schemes.

Side-channel and fault attacks. For signature schemes, the secrecy of the private key is critical. Care must be taken to protect implementations against attacks, such as side-channel attacks or fault attacks [17, 18, 19, 20, 21]. A cryptographic device may leak critical information with side-channel analysis or attacks that allow internal data or keying material to be extracted without breaking the cryptographic primitives.

4. Functions and Addressing

4.1 Hash Functions and Pseudorandom Functions

The specification of SLH-DSA makes use of six functions — $\mathbf{PRF}_{msg}$, $\mathbf{H}_{msg}$, $\mathbf{PRF}$, $\mathbf{T}_\ell$, $\mathbf{H}$, and $\mathbf{F}$ — that are all implemented using hash functions or XOFs with fixed output lengths. The inputs and output of each function are byte strings. In the following definitions, $\mathbb{B} = \{0, \dots, 255\}$ denotes the set of all bytes, $\mathbb{B}^n$ denotes the set of byte strings of length n bytes, and $\mathbb{B}^*$ denotes the set of all byte strings. The $\mathbf{ADRS}$ input is described in Section 4.2.

- $\mathbf{PRF}_{msg}(\mathbf{SK}.\mathrm{prf}, opt\_rand, M)$ $(\mathbb{B}^n \times \mathbb{B}^n \times \mathbb{B}^* \to \mathbb{B}^n)$ is a pseudorandom function (PRF) that generates the randomizer (R) for the randomized hashing of the message to be signed.

- $\mathbf{H}_{msg}(R, \mathbf{PK}.\mathrm{seed}, \mathbf{PK}.\mathrm{root}, M)$ $(\mathbb{B}^n \times \mathbb{B}^n \times \mathbb{B}^n \times \mathbb{B}^* \to \mathbb{B}^m)$ is used to generate the digest of the message to be signed.

- $\mathbf{PRF}(\mathbf{PK}.\mathrm{seed}, \mathbf{SK}.\mathrm{seed}, \mathbf{ADRS})$ $(\mathbb{B}^n \times \mathbb{B}^n \times \mathbb{B}^{32} \to \mathbb{B}^n)$ is a PRF that is used to generate the secret values in WOTS$^+$ and FORS private keys.

- $\mathbf{T}_\ell(\mathbf{PK}.\mathrm{seed}, \mathbf{ADRS}, M_\ell)$ $(\mathbb{B}^n \times \mathbb{B}^{32} \times \mathbb{B}^{\ell n} \to \mathbb{B}^n)$ is a hash function that maps an ℓn-byte message to an n-byte message.

- $\mathbf{H}(\mathbf{PK}.\mathrm{seed}, \mathbf{ADRS}, M_2)$ $(\mathbb{B}^n \times \mathbb{B}^{32} \times \mathbb{B}^{2n} \to \mathbb{B}^n)$ is a special case of $\mathbf{T}_\ell$ that takes a $2n$-byte message as input.

- $\mathbf{F}(\mathbf{PK}.\mathrm{seed}, \mathbf{ADRS}, M_1)$ $(\mathbb{B}^n \times \mathbb{B}^{32} \times \mathbb{B}^n \to \mathbb{B}^n)$ is a hash function that takes an n-byte message as input and produces an n-byte output.

The specific instantiations for these functions differ for different parameter sets and are specified in Section 11.

4.2 Addresses

Four of the functions described in Section 4.1 take a 32-byte address (i.e., $\mathbf{ADRS}$) as input. An $\mathbf{ADRS}$ consists of public values that indicate the position of the value being computed by the function. A different $\mathbf{ADRS}$ value is used for each call to each function. In the case of $\mathbf{PRF}$, this is used to generate a large number of different secret values from a single seed. In the case of $\mathbf{T}_\ell$, $\mathbf{H}$, and $\mathbf{F}$, it is used to mitigate multi-target attacks. In the pseudocode, where addresses are passed as parameters, they may be passed either by reference or by value.

The structure of an $\mathbf{ADRS}$ conforms to word boundaries, with each word being 4 bytes long and values encoded as unsigned integers in big-endian byte order (see Figure 2). The first word of $\mathbf{ADRS}$ specifies the layer address, which is the height of an XMSS tree within the hypertree. Trees on the bottom layer have a height of zero, and the single XMSS tree at the top has a height of $d - 1$ (see Figure 1). The next three words of $\mathbf{ADRS}$ specify the tree address, which is the position of an XMSS tree within a layer of the hypertree. The leftmost XMSS tree in a layer has a tree address of zero, and the rightmost XMSS tree in layer L has a tree address of $2^{(d-1-L)h'} - 1$. The next word is used to specify the type of the address, which differs depending on the use case.

There are seven different types of address used in SLH-DSA, as described below.[5] The address type determines how the final 12 bytes of the address are to be interpreted. The algorithms in this standard are written based on the assumption that whenever the type in an $\mathbf{ADRS}$ is changed, the final 12 bytes of the address are initialized to zero.

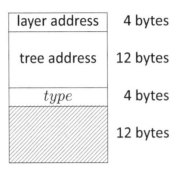

Figure 2. Address ($\mathbf{ADRS}$)

The type is set to WOTS_HASH (i.e., $type = 0$) for a WOTS$^+$ hash address (see Figure 3), which is used when computing hash chains in WOTS$^+$. When type is WOTS_HASH, the next word encodes the key pair address, which is the index of the WOTS$^+$ key pair within the XMSS tree specified by the layer and tree addresses, with the leftmost WOTS$^+$ key having an index of zero and the rightmost WOTS$^+$ key having an index of $2^{h'} - 1$. Next is the chain address, which encodes the index of the chain within WOTS$^+$, followed by the hash address, which encodes the address of the hash function within the chain.

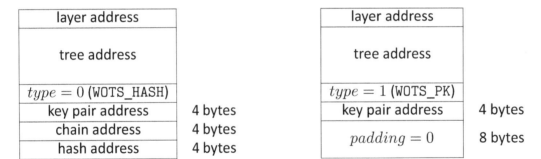

Figure 3. WOTS$^+$ hash address   Figure 4. WOTS$^+$ public-key compression address

The type is set to WOTS_PK (i.e., $type = 1$) when compressing WOTS$^+$ public keys (see Figure 4). As when the type is WOTS_HASH, the next word encodes the index of the WOTS$^+$ key pair within the XMSS tree specified by the layer and tree addresses. The remaining two words of $\mathbf{ADRS}$ are not needed and are set to zero.

The type is set to TREE (i.e., $type = 2$) when computing the hashes within the XMSS tree (see Figure 5). For this type of address, the next word is always set to zero. The following word

---

[5]The *type* word will have a value of 0, 1, 2, 3, 4, 5, or 6. In order to improve readability, these values will be referred to in this standard by the constants WOTS_HASH, WOTS_PK, TREE, FORS_TREE, FORS_ROOTS, WOTS_PRF, and FORS_PRF, respectively.

encodes the height of the node within the tree that is being computed, and the final word encodes the index of the node at that height.

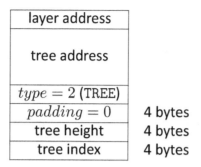

Figure 5. Hash tree address

The type is set to FORS_TREE (i.e., $type = 3$) when computing hashes within the FORS tree (see Figure 6). The next word is the key pair address, which encodes the FORS key that is used. The value is the same as the key pair address for the WOTS$^+$ key used to sign the FORS key (see Figure 3 and Figure 4). The next two words — the tree height and tree index — encode the node within the FORS tree that is being computed. The tree height starts with zero for the leaf nodes. The tree index is counted continuously across the k different FORS trees. The leftmost node in the leftmost tree has an index of zero, and the rightmost node in the rightmost tree at level j has an index of $k \cdot 2^{(a-j)} - 1$, where a is the height of the tree.

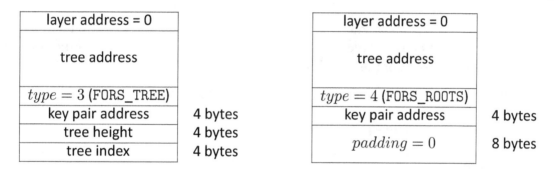

Figure 6. FORS tree address   **Figure 7. FORS tree roots compression address**

The type is set to FORS_ROOTS (i.e., $type = 4$) when compressing the k FORS tree roots (see Figure 7). The next word is the key pair address, which has the same meaning as it does in the FORS_TREE address. The remaining two words of **ADRS** are not needed and are set to zero.

The type is set to WOTS_PRF (i.e., $type = 5$) when generating secret values for WOTS$^+$ keys (see Figure 8). The values for the other words in the address are set to the same values as for the WOTS_HASH address (Figure 3) used for the chain. The hash address is always set to zero.

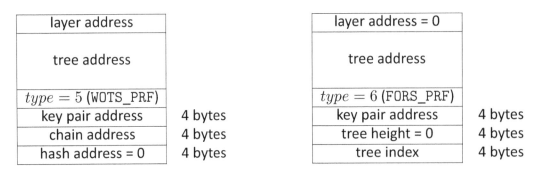

Figure 8. WOTS$^+$ key generation address   **Figure 9. FORS key generation address**

The type is set to FORS_PRF (i.e., $type = 6$) when generating secret values for FORS keys (see Figure 9). The values for the other words in the address are set to the same values as for the FORS_TREE address (Figure 6) used for the same leaf node.

4.3 Member Functions

The algorithms in this standard make use of member functions. If a complex data structure (e.g., an **ADRS**) contains a component X, then **ADRS**.getX() returns the value of X, and **ADRS**.setX(Y) sets the component X in **ADRS** to the value held by Y. If a data structure s contains multiple instances of X, then s.getX(i) returns the value of the ith instance of X in s. For example, if s is a FORS signature (Figure 14), then s.getAUTH(i) returns the authentication path for the ith tree.

Whenever the *type* in an address changes, the final 12 bytes of the address are initialized to zero. The member function **ADRS**.setTypeAndClear(Y) for addresses sets the *type* of the **ADRS** to Y and sets the final 12 bytes of the **ADRS** to zero.[6]

Table 1 shows alternative notation for each of the member functions that operates on addresses.

Table 1. Member functions for addresses

Member function	Expanded notation
ADRS.setLayerAddress(l)	$\textbf{ADRS} \leftarrow \text{toByte}(l, 4) \parallel \textbf{ADRS}[4:32]$
ADRS.setTreeAddress(t)	$\textbf{ADRS} \leftarrow \textbf{ADRS}[0:4] \parallel \text{toByte}(t, 12) \parallel \textbf{ADRS}[16:32]$
ADRS.setTypeAndClear(Y)	$\textbf{ADRS} \leftarrow \textbf{ADRS}[0:16] \parallel \text{toByte}(Y, 4) \parallel \text{toByte}(0, 12)$
ADRS.setKeyPairAddress(i)	$\textbf{ADRS} \leftarrow \textbf{ADRS}[0:20] \parallel \text{toByte}(i, 4) \parallel \textbf{ADRS}[24:32]$
ADRS.setChainAddress(i) **ADRS**.setTreeHeight(i)	$\textbf{ADRS} \leftarrow \textbf{ADRS}[0:24] \parallel \text{toByte}(i, 4) \parallel \textbf{ADRS}[28:32]$
ADRS.setHashAddress(i) **ADRS**.setTreeIndex(i)	$\textbf{ADRS} \leftarrow \textbf{ADRS}[0:28] \parallel \text{toByte}(i, 4)$
$i \leftarrow$ **ADRS**.getKeyPairAddress()	$i \leftarrow \text{toInt}(\textbf{ADRS}[20:24], 4)$
$i \leftarrow$ **ADRS**.getTreeIndex()	$i \leftarrow \text{toInt}(\textbf{ADRS}[28:32], 4)$

---

[6]As noted in Section 4.2, the type (Y) is an integer. However, in the pseudocode, the constants WOTS_HASH, WOTS_PK, TREE, FORS_TREE, FORS_ROOTS, WOTS_PRF, and FORS_PRF are used in order to improve readability.

4.4 Arrays, Byte Strings, and Integers

If X is a byte string of length n, then $X[i]$ for $i \in \{0, \ldots, n-1\}$ will refer to the ith element in the string X. If X is an array of m n-byte strings, then $X[i]$ for $i \in \{0, \ldots, m-1\}$ will refer to the ith n-byte string in X, and X will refer to the $m \cdot n$-byte string $X[0] \parallel X[1] \parallel \ldots X[m-1]$.

A byte string may be interpreted as the big-endian representation of an integer. In such cases, a byte string X of length n is converted to the integer

$$X[0] \cdot 256^{n-1} + X[1] \cdot 256^{n-2} + \ldots X[n-2] \cdot 256 + X[n-1].$$

Similarly, an integer x may be converted to a byte string of length n by finding coefficients $x_0, x_1, \ldots x_{n-2}, x_{n-1} \in \{0, \ldots, 255\}$ such that

$$x = x_0 \cdot 256^{n-1} + x_1 \cdot 256^{n-2} + \ldots x_{n-2} \cdot 256 + x_{n-1}$$

and then setting the byte string to be $x_0 x_1 \ldots x_{n-2} x_{n-1}$.

Algorithm 2 is a function that converts a byte string X of length n to an integer, and Algorithm 3 is a function that converts an integer x to a byte string of length n.

Algorithm 2 toInt(X, n)

Converts a byte string to an integer.

Input: n-byte string X.
Output: Integer value of X.

1: $total \leftarrow 0$

2: **for** i **from** 0 **to** $n - 1$ **do**
3:      $total \leftarrow 256 \cdot total + X[i]$
4: **end for**
5: **return** $total$

Algorithm 3 toByte(x, n)

Converts an integer to a byte string.

Input: Integer x, string length n.
Output: Byte string of length n containing binary representation of x in big-endian byte-order.

1: $total \leftarrow x$

2: **for** i **from** 0 **to** $n - 1$ **do**
3:      $S[n - 1 - i] \leftarrow total \bmod 256$ ▷ least significant 8 bits of $total$
4:      $total \leftarrow total \gg 8$
5: **end for**
6: **return** S

For the WOTS$^+$ and FORS schemes, the messages to be signed need to be split into a sequence of b-bit strings, where each b-bit string is interpreted as an integer between 0 and $2^b - 1$.[7] This is the equivalent of creating the base-2^b representation of the message. The base_2^b function (Algorithm 4) takes a byte string X, a bit string length b, and an output length out_len as input and returns an array of base-2^b integers that represent the first $out\_len \cdot b$ bits of X if the individual bytes in X are encoded as 8-bit strings in big-endian bit order. X must be at least $\lceil out\_len \cdot b/8 \rceil$ bytes in length. As the value of $bits$ will never exceed $b + 7$, a $b + 7$-bit unsigned integer is sufficient to store $total$ (i.e., $total$ may be stored as a 32-bit unsigned integer).

Algorithm 4 base_2^b(X, b, out_len)

Computes the base 2^b representation of X.

Input: Byte string X of length at least $\left\lceil \frac{out\_len \cdot b}{8} \right\rceil$, integer b, output length out_len.

Output: Array of out_len integers in the range $[0, \ldots, 2^b - 1]$.

1: $in \leftarrow 0$
2: $bits \leftarrow 0$
3: $total \leftarrow 0$

4: **for** out **from** 0 **to** $out\_len - 1$ **do**
5:     **while** $bits < b$ **do**
6:         $total \leftarrow (total \ll 8) + X[in]$
7:         $in \leftarrow in + 1$
8:         $bits \leftarrow bits + 8$
9:     **end while**
10:     $bits \leftarrow bits - b$
11:     $baseb[out] \leftarrow (total \gg bits) \bmod 2^b$
12: **end for**
13: **return** $baseb$

---

[7] b will be the value of lg_w when the base_2^b function is used in WOTS$^+$, and b will be the value of a when the base_2^b function is used in FORS. For the parameter sets in this standard, lg_w is 4, and a is 6, 8, 9, 12, or 14.

5. Winternitz One-Time Signature Plus Scheme

This section describes the WOTS$^+$ one-time signature scheme that is a component of SLH-DSA.

WOTS$^+$ uses two parameters. The security parameter n is the length in bytes of the messages that may be signed as well as the length of the private key elements, public key elements, and signature elements. For the parameter sets specified in this standard, n may be 16, 24, or 32 (see Table 2). The second parameter lg_w indicates the number of bits that are encoded by each hash chain that is used.[8] lg_w is 4 for all parameter sets in this standard. These parameters are used to compute four additional values:

$$w = 2^{lg_w} \tag{5.1}$$

$$len_1 = \left\lceil \frac{8n}{lg_w} \right\rceil \tag{5.2}$$

$$len_2 = \left\lfloor \frac{\log_2(len_1 \cdot (w-1))}{lg_w} \right\rfloor + 1 \tag{5.3}$$

$$len = len_1 + len_2 \tag{5.4}$$

When $lg_w = 4$, $w = 16$, $len_1 = 2n$, $len_2 = 3$, and $len = 2n + 3$.

A WOTS$^+$ private key consists of len secret values of length n. In SLH-DSA, these are all generated from an n-byte seed $\mathbf{SK}$.seed using a PRF. Chains of length w are then created from the secret values using a chaining function, and the end values from each of the chains are public values. The WOTS$^+$ public key is computed as the hash of these public values. In order to create a signature, the $8n$-bit message is first converted into an array of len_1 base-w integers. A checksum is then computed for this string, and the checksum is converted into an array of len_2 base-w integers. The signature consists of the appropriate entries from the chains for each of the integers in the message and checksum arrays.

The WOTS$^+$ functions make use of two helper functions: base_2^b and chain. The base_2^b function (Section 4.4) is used to break the message to be signed and the checksum value into arrays of base-w integers. The chain function (Algorithm 5) is used to compute the hash chains.

The chain function takes an n-byte string X and integers s and i (where $i + s < w$) as input and returns the result of iterating a hash function $\mathbf{F}$ on the input s times, starting from an index of i.[9] The chain function also requires as input $\mathbf{PK}$.seed, which is part of the SLH-DSA public key, and an address $\mathbf{ADRS}$. The $type$ in $\mathbf{ADRS}$ must be set to WOTS_HASH, and the layer address, tree address, key pair address, and chain address must be set to the address of the chain being computed. The chain function updates the hash address in $\mathbf{ADRS}$ with each iteration to specify the current position in the chain prior to $\mathbf{ADRS}$'s use in $\mathbf{F}$.

---

[8]In [10], the Winternitz parameter w is used as the second WOTS$^+$ parameter, where w indicates the length of the
 hash chains that are used. This standard uses the parameter $lg_w = \log_2(w)$ to simplify notation.
[9]A start index of 0 indicates the beginning of the chain.

17

Algorithm 5 chain($X, i, s, \mathbf{PK}.\text{seed}, \mathbf{ADRS}$)

Chaining function used in WOTS$^+$.

Input: Input string X, start index i, number of steps s, public seed $\mathbf{PK}.\text{seed}$, address $\mathbf{ADRS}$.
Output: Value of $\mathbf{F}$ iterated s times on X.

1: $tmp \leftarrow X$

2: **for** j **from** i **to** $i + s - 1$ **do**
3:     $\mathbf{ADRS}.\text{setHashAddress}(j)$
4:     $tmp \leftarrow \mathbf{F}(\mathbf{PK}.\text{seed}, \mathbf{ADRS}, tmp)$
5: **end for**
6: **return** tmp

5.1 WOTS$^+$ Public-Key Generation

The wots_pkGen function (Algorithm 6) generates WOTS$^+$ public keys. It takes $\mathbf{SK}.\text{seed}$ and $\mathbf{PK}.\text{seed}$ from the SLH-DSA private key and an address as input. The *type* in the address $\mathbf{ADRS}$ must be set to WOTS_HASH, and the layer address, tree address, and key pair address must encode the address of the WOTS$^+$ public key to be generated.

Lines 4 through 9 in Algorithm 6 generate the public values, as described in Section 5. For each of the len public values, the corresponding secret value is generated in lines 5 and 6, and the chain function is called to compute the end value of the chain of length w. Once the len public values are computed, they are compressed into a single n-byte value in lines 10 through 13.

Algorithm 6 wots_pkGen($\mathbf{SK}.\text{seed}, \mathbf{PK}.\text{seed}, \mathbf{ADRS}$)

Generates a WOTS$^+$ public key.

Input: Secret seed $\mathbf{SK}.\text{seed}$, public seed $\mathbf{PK}.\text{seed}$, address $\mathbf{ADRS}$.
Output: WOTS$^+$ public key pk.

1: skADRS $\leftarrow \mathbf{ADRS}$ ▷ copy address to create key generation key address
2: skADRS.setTypeAndClear(WOTS_PRF)
3: skADRS.setKeyPairAddress($\mathbf{ADRS}.\text{getKeyPairAddress}()$)
4: **for** i **from** 0 **to** $len - 1$ **do**
5:     skADRS.setChainAddress(i)
6:     $sk \leftarrow \mathbf{PRF}(\mathbf{PK}.\text{seed}, \mathbf{SK}.\text{seed}, \text{skADRS})$ ▷ compute secret value for chain i
7:     $\mathbf{ADRS}.\text{setChainAddress}(i)$
8:     $tmp[i] \leftarrow \text{chain}(sk, 0, w - 1, \mathbf{PK}.\text{seed}, \mathbf{ADRS})$ ▷ compute public value for chain i
9: **end for**
10: wotspkADRS $\leftarrow \mathbf{ADRS}$ ▷ copy address to create WOTS$^+$ public key address
11: wotspkADRS.setTypeAndClear(WOTS_PK)
12: wotspkADRS.setKeyPairAddress($\mathbf{ADRS}.\text{getKeyPairAddress}()$)
13: $pk \leftarrow \mathbf{T}_{len}(\mathbf{PK}.\text{seed}, \text{wotspkADRS}, tmp)$ ▷ compress public key
14: **return** pk

5.2 WOTS$^+$ Signature Generation

A WOTS$^+$ signature is an array of len byte strings of length n, as shown in Figure 10. The wots_sign function (Algorithm 7) generates the signature by converting the n-byte message M[10] into an array of len_1 base-w integers (line 2). A checksum is computed over M (lines 3 through 5). The checksum is converted to a byte string, which is then converted into an array of len_2 base-w integers (lines 6 and 7). The len_2 integers that represent the checksum are appended to the len_1 integers that represent the message (line 7).[11] For each of the len base-w integers, the signature consists of the corresponding node in one of the hash chains. For each of these integers, lines 12 and 13 compute the secret value for the hash chain, and lines 14 and 15 compute the node in the hash chain that corresponds to the integer. The selected nodes are concatenated to form the WOTS$^+$ signature.

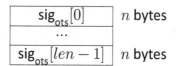

Figure 10. WOTS$^+$ signature data format

In addition to the n-byte message to be signed, wots_sign takes **SK**.seed and **PK**.seed from the SLH-DSA private key and an address as input. The *type* in the address **ADRS** must be set to WOTS_HASH, and the layer address, tree address, and key pair address must encode the address of the WOTS$^+$ key that is used to sign the message.

---

[10] In SLH-DSA, the message M that is signed using WOTS$^+$ is either an XMSS public key or a FORS public key.

[11] In the case that $lg_w = 4$, the n-byte message is converted into an array of $2n$ base-16 integers (i.e., hexadecimal digits). The checksum is encoded as two bytes with the least significant four bits being zeros, and the most significant 12 bits are appended to the message as an array of three base-16 integers.

Algorithm 7 wots_sign(M, $\mathbf{SK}$.seed, $\mathbf{PK}$.seed, $\mathbf{ADRS}$)

Generates a WOTS$^+$ signature on an n-byte message.

Input: Message M, secret seed $\mathbf{SK}$.seed, public seed $\mathbf{PK}$.seed, address $\mathbf{ADRS}$.
Output: WOTS$^+$ signature sig.

1: $csum \leftarrow 0$

2: $msg \leftarrow \text{base\_2}^b(M, lg_w, len_1)$ $\triangleright$ convert message to base w

3: **for** i **from** 0 **to** $len_1 - 1$ **do** $\triangleright$ compute checksum

4:      $csum \leftarrow csum + w - 1 - msg[i]$

5: **end for**

6: $csum \leftarrow csum \ll ((8 - ((len_2 \cdot lg_w) \bmod 8)) \bmod 8)$ $\triangleright$ for $lg_w = 4$, left shift by 4

7: $msg \leftarrow msg \parallel \text{base\_2}^b\left(\text{toByte}\left(csum, \left\lceil \frac{len_2 \cdot lg_w}{8} \right\rceil\right), lg_w, len_2\right)$ $\triangleright$ convert to base w

8: skADRS $\leftarrow \mathbf{ADRS}$ $\triangleright$ copy address to create key generation key address

9: skADRS.setTypeAndClear(WOTS_PRF)

10: skADRS.setKeyPairAddress(ADRS.getKeyPairAddress())

11: **for** i **from** 0 **to** $len - 1$ **do**

12:      skADRS.setChainAddress(i)

13:      $sk \leftarrow \mathbf{PRF}(\mathbf{PK}.\text{seed}, \mathbf{SK}.\text{seed}, \text{skADRS})$ $\triangleright$ compute chain i secret value

14:      $\mathbf{ADRS}.\text{setChainAddress}(i)$

15:      $sig[i] \leftarrow \text{chain}(sk, 0, msg[i], \mathbf{PK}.\text{seed}, \mathbf{ADRS})$ $\triangleright$ compute chain i signature value

16: **end for**

17: **return** sig

5.3 Computing a WOTS$^+$ Public Key From a Signature

Verifying a WOTS$^+$ signature involves computing a public-key value from a message and signature value. Verification succeeds if the correct public-key value is computed, which is determined by using the computed public-key value along with other information to compute a candidate $\mathbf{PK}$.root value and then comparing that value to the known value of $\mathbf{PK}$.root from the SLH-DSA public key. This section describes wots_pkFromSig (Algorithm 8), a function that computes a candidate WOTS$^+$ public key from a WOTS$^+$ signature and corresponding message.

In addition to an n-byte message M and a $len \cdot n$-byte signature sig, which is interpreted as an array of len n-byte strings, the wots_pkFromSig function takes $\mathbf{PK}$.seed from the SLH-DSA public key and an address as input. The *type* of the address $\mathbf{ADRS}$ must be set to WOTS_HASH, and the layer address, tree address, and key pair address must encode the address of the WOTS$^+$ key that was used to sign the message.

Lines 1 through 7 of wots_pkFromSig are the same as lines 1 through 7 of wots_sign (Algorithm 7). Lines 8 through 11 of wots_pkFromSig compute the end nodes for each of the chains using the signature value as the starting point and the message value to determine the number of iterations that need to be performed to get to the end node. Finally, as with lines 10 through 13 of Algorithm 6, the computed public-key values are compressed in lines 12 through 15.

Algorithm 8 wots_pkFromSig(sig, M, $\mathbf{PK}$.seed, $\mathbf{ADRS}$)

Computes a WOTS$^+$ public key from a message and its signature.

Input: WOTS$^+$ signature sig, message M, public seed $\mathbf{PK}$.seed, address $\mathbf{ADRS}$.
Output: WOTS$^+$ public key pk_{sig} derived from sig.

1: $csum \leftarrow 0$

2: $msg \leftarrow \text{base\_2}^b(M, lg_w, len_1)$ $\qquad\qquad\qquad\qquad$ ▷ convert message to base w

3: **for** i from 0 to $len_1 - 1$ **do** $\qquad\qquad\qquad\qquad$ ▷ compute checksum
4: $\quad csum \leftarrow csum + w - 1 - msg[i]$
5: **end for**

6: $csum \leftarrow csum \ll ((8 - ((len_2 \cdot lg_w) \bmod 8)) \bmod 8)$ $\quad$ ▷ for $lg_w = 4$, left shift by 4
7: $msg \leftarrow msg \parallel \text{base\_2}^b\left(\text{toByte}\left(csum, \lceil \frac{len_2 \cdot lg_w}{8} \rceil\right), lg_w, len_2\right)$ $\quad$ ▷ convert to base w

8: **for** i from 0 to $len - 1$ **do**
9: $\quad \mathbf{ADRS}.\text{setChainAddress}(i)$
10: $\quad tmp[i] \leftarrow \text{chain}(sig[i], msg[i], w - 1 - msg[i], \mathbf{PK}.\text{seed}, \mathbf{ADRS})$
11: **end for**
12: wotspkADRS $\leftarrow \mathbf{ADRS}$ $\qquad\qquad$ ▷ copy address to create WOTS$^+$ public key address
13: wotspkADRS.setTypeAndClear(WOTS_PK)
14: wotspkADRS.setKeyPairAddress($\mathbf{ADRS}$.getKeyPairAddress())
15: $pk_{sig} \leftarrow \mathbf{T}_{len}(\mathbf{PK}.\text{seed}, \text{wotspkADRS}, tmp)$
16: **return** pk_{sig}

---

6.   eXtended Merkle Signature Scheme (XMSS)

XMSS extends the WOTS$^+$ signature scheme into one that can sign multiple messages. A Merkle tree [13] of height h' is used to allow $2^{h'}$ WOTS$^+$ public keys to be authenticated using a single n-byte XMSS public key, which is the root of the Merkle tree.[12] As each WOTS$^+$ key may be used to sign one message, the XMSS key may be used to sign $2^{h'}$ messages.

An XMSS signature is $(h' + len) \cdot n$ bytes in length and consists of a WOTS$^+$ signature and an authentication path (see Figure 11). The authentication path is an array of nodes from the Merkle tree — one from each level of the tree, except for the root — that allows the verifier to compute the root of the tree when used in conjunction with the WOTS$^+$ public key that can be computed from the WOTS$^+$ signature.

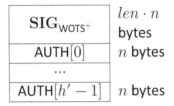

Figure 11. XMSS signature data format

6.1   Generating a Merkle Hash Tree

The xmss_node function (Algorithm 9) computes the nodes of an XMSS tree. The xmss_node function takes as input **SK**.seed and **PK**.seed from the SLH-DSA private key; a target node index i, which is the index of the node being computed; a target node height z, which is the height within the Merkle tree of the node being computed; and an address. The address **ADRS** must have the layer address and tree address set to the XMSS tree within which the node is being computed. The target node height and index must satisfy $z \leq h'$ and $i < 2^{(h'-z)}$.

Each node in an XMSS tree is the root of a subtree, and Algorithm 9 computes the root of the subtree recursively. If the subtree consists of a single leaf node, then the function simply returns the value of the node's WOTS$^+$ public key (lines 2 through 4). Otherwise, the function computes the roots of the left subtree (line 6) and right subtree (line 7) and hashes them together (lines 8 through 11).

---

[12]The Merkle tree formed from the $2^{h'}$ WOTS$^+$ keys of an XMSS key is referred to in this standard as an XMSS tree.

Algorithm 9 xmss_node($\mathbf{SK}$.seed, i, z, $\mathbf{PK}$.seed, $\mathbf{ADRS}$)

Computes the root of a Merkle subtree of WOTS$^+$ public keys.

Input: Secret seed $\mathbf{SK}$.seed, target node index i, target node height z, public seed $\mathbf{PK}$.seed, address $\mathbf{ADRS}$.

Output: n-byte root $node$.

 1: **if** $z = 0$ **then**
 2:     $\mathbf{ADRS}$.setTypeAndClear(WOTS_HASH)
 3:     $\mathbf{ADRS}$.setKeyPairAddress(i)
 4:     $node \leftarrow$ wots_pkGen($\mathbf{SK}$.seed, $\mathbf{PK}$.seed, $\mathbf{ADRS}$)
 5: **else**
 6:     $lnode \leftarrow$ xmss_node($\mathbf{SK}$.seed, $2i$, $z - 1$, $\mathbf{PK}$.seed, $\mathbf{ADRS}$)
 7:     $rnode \leftarrow$ xmss_node($\mathbf{SK}$.seed, $2i + 1$, $z - 1$, $\mathbf{PK}$.seed, $\mathbf{ADRS}$)
 8:     $\mathbf{ADRS}$.setTypeAndClear(TREE)
 9:     $\mathbf{ADRS}$.setTreeHeight(z)
10:     $\mathbf{ADRS}$.setTreeIndex(i)
11:     $node \leftarrow \mathbf{H}(\mathbf{PK}$.seed, $\mathbf{ADRS}$, $lnode \parallel rnode)$
12: **end if**
13: **return** $node$

6.2 Generating an XMSS Signature

The xmss_sign function (Algorithm 10) creates an XMSS signature on an n-byte message M[13] by creating an authentication path (lines 1 through 4) and signing M with the appropriate WOTS$^+$ key (lines 5 through 7). In addition to M, xmss_sign takes $\mathbf{SK}$.seed and $\mathbf{PK}$.seed from the SLH-DSA private key, an address, and an index as input. The address $\mathbf{ADRS}$ must have the layer address and tree address set to the XMSS key that is being used to sign the message, and the index idx must be the index of the WOTS$^+$ key within the XMSS tree that will be used to sign the message.

The authentication path consists of the sibling nodes of each node that is on the path from the WOTS$^+$ key used to the root. For example, in Figure 12, if the message is signed with K_2, then K_2, $n_{1,1}$, and $n_{2,0}$ are the on path nodes, and the authentication path consists of K_3, $n_{1,0}$, and $n_{2,1}$. In line 2 of Algorithm 10, $\lfloor idx/2^j \rfloor$ is the on path node, and $\lfloor idx/2^j \rfloor \oplus 1$ is the authentication path node. Line 3 computes the value of the authentication path node.

---

[13] In SLH-DSA, the message M that is signed using XMSS is either an XMSS public key or a FORS public key.

Algorithm 10 xmss_sign(M, **SK**.seed, idx, **PK**.seed, **ADRS**)

Generates an XMSS signature.

Input: n-byte message M, secret seed **SK**.seed, index idx, public seed **PK**.seed,
    address **ADRS**.

Output: XMSS signature $\text{SIG}_{XMSS} = (sig \parallel \text{AUTH})$.

1: **for** j from 0 **to** $h' - 1$ **do**   ▷ build authentication path
2:     $k \leftarrow \lfloor idx/2^j \rfloor \oplus 1$
3:     $\text{AUTH}[j] \leftarrow$ xmss_node(**SK**.seed, k, j, **PK**.seed, **ADRS**)
4: **end for**

5: **ADRS**.setTypeAndClear(WOTS_HASH)
6: **ADRS**.setKeyPairAddress(idx)
7: $sig \leftarrow$ wots_sign(M, **SK**.seed, **PK**.seed, **ADRS**)
8: $\text{SIG}_{XMSS} \leftarrow sig \parallel \text{AUTH}$
9: **return** SIG_{XMSS}

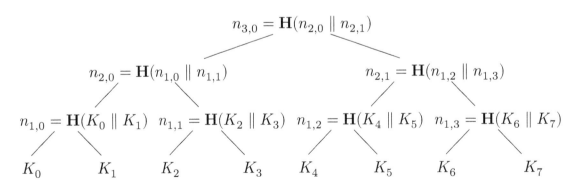

Figure 12. Merkle hash tree

6.3 Computing an XMSS Public Key From a Signature

Verifying an XMSS signature involves computing a public-key value from a message and a signature value. Verification succeeds if the correct public-key value is computed, which is determined by using the computed public-key value along with other information to compute a candidate $\mathbf{PK}$.root value and then comparing that value to the known value of $\mathbf{PK}$.root from the SLH-DSA public key. This section describes xmss_pkFromSig (Algorithm 11), a function that computes a candidate XMSS public key from an XMSS signature and corresponding message.

In addition to an n-byte message M and an $(len+h')\cdot n$-byte signature SIG_{XMSS}, xmss_pkFromSig takes $\mathbf{PK}$.seed from the SLH-DSA public key, an address, and an index as input. The address $\mathbf{ADRS}$ must be set to the layer address and tree address of the XMSS key that was used to sign the message, and the index idx must be the index of the WOTS$^+$ key within the XMSS tree that was used to sign the message.

Algorithm 11 begins by computing the WOTS$^+$ public key in lines 1 through 5. The root is then computed in lines 6 through 18. Starting with the leaf node (i.e., the WOTS$^+$ public key), a node at each level of the tree is computed by hashing together the node computed in the previous iteration with the corresponding authentication path node. In lines 12 and 15, AUTH is interpreted as an array of h' n-byte strings.

Algorithm 11 xmss_pkFromSig(idx, SIG_{XMSS}, M, $\mathbf{PK}$.seed, $\mathbf{ADRS}$)

Computes an XMSS public key from an XMSS signature.

Input: Index idx, XMSS signature $\text{SIG}_{XMSS} = (sig \parallel \text{AUTH})$, n-byte message M,
    public seed $\mathbf{PK}$.seed, address $\mathbf{ADRS}$.
Output: n-byte root value $node[0]$.

  1: $\mathbf{ADRS}$.setTypeAndClear(WOTS_HASH)   ▷ compute WOTS$^+$ pk from WOTS$^+$ sig
  2: $\mathbf{ADRS}$.setKeyPairAddress(idx)
  3: $sig \leftarrow \text{SIG}_{XMSS}$.getWOTSSig()   ▷ $\text{SIG}_{XMSS}[0 : len \cdot n]$
  4: $\text{AUTH} \leftarrow \text{SIG}_{XMSS}$.getXMSSAUTH()   ▷ $\text{SIG}_{XMSS}[len \cdot n : (len + h') \cdot n]$
  5: $node[0] \leftarrow$ wots_pkFromSig(sig, M, $\mathbf{PK}$.seed, $\mathbf{ADRS}$)

  6: $\mathbf{ADRS}$.setTypeAndClear(TREE)   ▷ compute root from WOTS$^+$ pk and AUTH
  7: $\mathbf{ADRS}$.setTreeIndex(idx)
  8: **for** k **from** 0 **to** $h' - 1$ **do**
  9:     $\mathbf{ADRS}$.setTreeHeight($k + 1$)
10:     **if** $\lfloor idx/2^k \rfloor$ is even **then**
11:         $\mathbf{ADRS}$.setTreeIndex($\mathbf{ADRS}$.getTreeIndex()$/2$)
12:         $node[1] \leftarrow \mathbf{H}(\mathbf{PK}$.seed, $\mathbf{ADRS}$, $node[0] \parallel \text{AUTH}[k])$
13:     **else**
14:         $\mathbf{ADRS}$.setTreeIndex(($\mathbf{ADRS}$.getTreeIndex() $- 1)/2$)
15:         $node[1] \leftarrow \mathbf{H}(\mathbf{PK}$.seed, $\mathbf{ADRS}$, $\text{AUTH}[k] \parallel node[0])$
16:     **end if**
17:     $node[0] \leftarrow node[1]$
18: **end for**
19: **return** $node[0]$

---

7. The SLH-DSA Hypertree

SLH-DSA requires a very large number of WOTS$^+$ keys to sign FORS public keys. As it would not be feasible for the parameter sets in this standard to have a single XMSS key with so many WOTS$^+$ keys, SLH-DSA uses a hypertree to sign the FORS keys. As depicted in Figure 1, a hypertree is a tree of XMSS trees. The XMSS keys at the lowest layer are used to sign FORS public keys (Section 8), and the XMSS keys at every other layer are used to sign the XMSS public keys at the layer below.

The hypertree has d layers of XMSS trees with each XMSS tree being a Merkle tree of height h', so the total height of the hypertree is $h = d \cdot h'$ (see Table 2). The top layer (layer $d - 1$) is a single XMSS tree, and the public key of this XMSS key pair (i.e., the root of the Merkle tree) is the public key of the hypertree (**PK**.root). The next layer down has $2^{h'}$ XMSS trees, and the public key of each of these XMSS keys is signed by one of the $2^{h'}$ WOTS$^+$ keys that is part of the top layer's XMSS key. The lowest layer has $2^{h-h'}$ XMSS trees, providing 2^h WOTS$^+$ keys to sign FORS keys.

7.1 Hypertree Signature Generation

A hypertree signature is $(h + d \cdot len) \cdot n$ bytes in length and consists of a sequence of d XMSS signatures, starting with a signature generated using an XMSS key at the lowest layer and ending with a signature generated using the XMSS key at the top layer (see Figure 13).

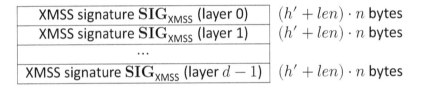

XMSS signature $\mathbf{SIG}_{\text{XMSS}}$ (layer 0)	$(h' + len) \cdot n$ bytes
XMSS signature $\mathbf{SIG}_{\text{XMSS}}$ (layer 1)	$(h' + len) \cdot n$ bytes
...	
XMSS signature $\mathbf{SIG}_{\text{XMSS}}$ (layer $d - 1$)	$(h' + len) \cdot n$ bytes

Figure 13. HT signature data format

In addition to the n-byte message M,[14] the ht_sign function (Algorithm 12) takes as input **SK**.seed and **PK**.seed from the SLH-DSA private key, the index of the XMSS tree at the lowest layer that will sign the message idx_{tree}, and the index of the WOTS$^+$ key within the XMSS tree that will sign the message idx_{leaf}.

Algorithm 12 begins in lines 1 through 3 by signing M with the specified XMSS key using the WOTS$^+$ key within that XMSS key specified by idx_{leaf}. The XMSS public key is obtained (line 5 or 14) for each successive layer and signed by the appropriate key at the next higher level (lines 7 through 11).

---

[14]In SLH-DSA, the message M that is provided to ht_sign is a FORS public key.

Algorithm 12 ht_sign(M, **SK**.seed, **PK**.seed, idx_{tree}, idx_{leaf})

Generates a hypertree signature.

Input: Message M, private seed **SK**.seed, public seed **PK**.seed, tree index idx_{tree}, leaf index idx_{leaf}.

Output: HT signature SIG_{HT}.

1: $\textbf{ADRS} \leftarrow \text{toByte}(0, 32)$

2: $\textbf{ADRS}.\text{setTreeAddress}(idx_{tree})$
3: $\text{SIG}_{tmp} \leftarrow \text{xmss\_sign}(M, \textbf{SK}.\text{seed}, idx_{leaf}, \textbf{PK}.\text{seed}, \textbf{ADRS})$
4: $\text{SIG}_{HT} \leftarrow \text{SIG}_{tmp}$
5: $root \leftarrow \text{xmss\_pkFromSig}(idx_{leaf}, \text{SIG}_{tmp}, M, \textbf{PK}.\text{seed}, \textbf{ADRS})$
6: **for** j **from** 1 **to** $d - 1$ **do**
7:      $idx_{leaf} \leftarrow idx_{tree} \bmod 2^{h'}$ $\triangleright$ h' least significant bits of idx_{tree}
8:      $idx_{tree} \leftarrow idx_{tree} \gg h'$ $\triangleright$ remove least significant h' bits from idx_{tree}
9:      $\textbf{ADRS}.\text{setLayerAddress}(j)$
10:      $\textbf{ADRS}.\text{setTreeAddress}(idx_{tree})$
11:      $\text{SIG}_{tmp} \leftarrow \text{xmss\_sign}(root, \textbf{SK}.\text{seed}, idx_{leaf}, \textbf{PK}.\text{seed}, \textbf{ADRS})$
12:      $\text{SIG}_{HT} \leftarrow \text{SIG}_{HT} \parallel \text{SIG}_{tmp}$
13:      **if** $j < d - 1$ **then**
14:          $root \leftarrow \text{xmss\_pkFromSig}(idx_{leaf}, \text{SIG}_{tmp}, root, \textbf{PK}.\text{seed}, \textbf{ADRS})$
15:      **end if**
16: **end for**
17: **return** SIG_{HT}

7.2   Hypertree Signature Verification

Hypertree signature verification works by making d calls to xmss_pkFromSig (Algorithm 11) and comparing the result to the public key of the hypertree.

In addition to the n-byte message M and the $(h + d \cdot len) \cdot n$-byte signature SIG_{HT}, ht_verify (Algorithm 13) takes as input $\mathbf{PK}$.seed and $\mathbf{PK}$.root from the SLH-DSA public key, the index of the XMSS tree at the lowest layer that signed the message idx_{tree}, and the index of the WOTS$^+$ key within the XMSS tree that signed the message idx_{leaf}.

At each layer, either the message M or the computed public key of the XMSS key at the lower layer is provided along with the appropriate XMSS signature to xmss_pkFromSig in order to obtain the layer's computed XMSS public key. If the computed XMSS public key of the top layer tree is the same as the known hypertree public key, $\mathbf{PK}$.root, then verification succeeds.

Algorithm 13 ht_verify(M, SIG_{HT}, $\mathbf{PK}$.seed, idx_{tree}, idx_{leaf}, $\mathbf{PK}$.root)

Verifies a hypertree signature.

Input: Message M, signature SIG_{HT}, public seed $\mathbf{PK}$.seed, tree index idx_{tree}, leaf index idx_{leaf}, HT public key $\mathbf{PK}$.root.

Output: Boolean.

1: $\mathbf{ADRS} \leftarrow$ toByte$(0, 32)$

2: $\mathbf{ADRS}$.setTreeAddress(idx_{tree})

3: $\text{SIG}_{tmp} \leftarrow \text{SIG}_{HT}$.getXMSSSignature$(0)$ $\triangleright \text{SIG}_{HT}[0 : (h' + len) \cdot n]$

4: $node \leftarrow$ xmss_pkFromSig$(idx_{leaf}, \text{SIG}_{tmp}, M, \mathbf{PK}.\text{seed}, \mathbf{ADRS})$

5: **for** j **from 1 to** $d - 1$ **do**

6:    $idx_{leaf} \leftarrow idx_{tree}$ mod $2^{h'}$ $\triangleright$ h' least significant bits of idx_{tree}

7:    $idx_{tree} \leftarrow idx_{tree} \gg h'$ $\triangleright$ remove least significant h' bits from idx_{tree}

8:    $\mathbf{ADRS}$.setLayerAddress(j)

9:    $\mathbf{ADRS}$.setTreeAddress(idx_{tree})

10:    $\text{SIG}_{tmp} \leftarrow \text{SIG}_{HT}$.getXMSSSignature$(j)$ $\triangleright \text{SIG}_{HT}[j \cdot (h' + len) \cdot n : (j+1)(h' + len) \cdot n]$

11:    $node \leftarrow$ xmss_pkFromSig$(idx_{leaf}, \text{SIG}_{tmp}, node, \mathbf{PK}.\text{seed}, \mathbf{ADRS})$

12: **end for**

13: **if** $node = \mathbf{PK}$.root **then**

14:    **return** true

15: **else**

16:    **return** false

17: **end if**

---

8.   Forest of Random Subsets (FORS)

FORS is a few-time signature scheme that is used to sign the digests of the actual messages. Unlike WOTS$^+$, for which forgeries become feasible if a key is used twice [22], the security of a FORS key degrades gradually as the number of signatures increases.

FORS uses two parameters: k and $t = 2^a$ (see Table 2). A FORS private key consists of k sets of t n-byte strings, all of which are pseudorandomly generated from the seed SK.seed. Each of the k sets is formed into a Merkle tree, and the roots of the trees are hashed together to form the FORS public key. A signature on a $k \cdot a$-bit message digest consists of k elements from the private key, one from each set selected using a bits of the message digest, along with the authentication paths for each of these elements (see Figure 14).

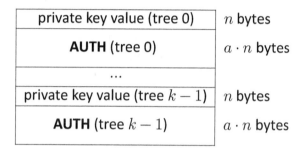

private key value (tree 0)	n bytes
AUTH (tree 0)	$a \cdot n$ bytes
...	
private key value (tree $k-1$)	n bytes
AUTH (tree $k-1$)	$a \cdot n$ bytes

Figure 14. FORS signature data format

8.1   Generating FORS Secret Values

The fors_skGen function (Algorithm 14) generates the n-byte strings of the FORS private key. The function takes SK.seed and PK.seed from the SLH-DSA private key, an address, and an index as input. The $type$ in the address ADRS must be set to FORS_TREE, and the tree address and key pair address must be set to the index of the WOTS$^+$ key within the XMSS tree that signs the FORS key. The layer address must be set to zero. The index idx is the index of the FORS secret value within the sets of FORS trees.

Algorithm 14 fors_skGen(SK.seed, PK.seed, ADRS, idx)

Generates a FORS private-key value.

Input: Secret seed SK.seed, public seed PK.seed, address ADRS, secret key index idx.
Output: n-byte FORS private-key value.

1: skADRS ← ADRS   ▷ copy address to create key generation address
2: skADRS.setTypeAndClear(FORS_PRF)
3: skADRS.setKeyPairAddress(ADRS.getKeyPairAddress())
4: skADRS.setTreeIndex(idx)
5: **return** PRF(PK.seed, SK.seed, skADRS)

8.2   Generating a Merkle Hash Tree

The fors_node function (Algorithm 15) computes the nodes of a Merkle tree. It is the same as xmss_node, except that the leaf nodes are the hashes of the FORS secret values instead of WOTS$^+$ public keys.

The fors_node function takes as input $\mathbf{SK}$.seed and $\mathbf{PK}$.seed from the SLH-DSA private key; a target node index i, which is the index of the node being computed; a target node height z, which is the height within the Merkle tree of the node being computed; and an address. The address $\mathbf{ADRS}$ must have the layer address set to zero (since the XMSS tree that signs a FORS key is always at layer 0), the tree address set to the XMSS tree that signs the FORS key, the $type$ set to FORS_TREE, and the key pair address set to the index of the WOTS$^+$ key within the XMSS tree that signs the FORS key. The target node height and index must satisfy $z \leq a$ and $i < k \cdot 2^{(a-z)}$.

Each node in the Merkle tree is the root of a subtree, and Algorithm 15 computes the root of a subtree recursively. If the subtree consists of a single leaf node, then the function simply returns a hash of the node's private n-byte string (lines 2 through 5). Otherwise, the function computes the roots of the left subtree (line 7) and right subtree (line 8) and hashes them together (lines 9 through 11).

Algorithm 15 fors_node($\mathbf{SK}$.seed, i, z, $\mathbf{PK}$.seed, $\mathbf{ADRS}$)

Computes the root of a Merkle subtree of FORS public values.

Input: Secret seed $\mathbf{SK}$.seed, target node index i, target node height z, public seed $\mathbf{PK}$.seed, address $\mathbf{ADRS}$.
Output: n-byte root $node$.

1: **if** $z = 0$ **then**
2:      $sk \leftarrow$ fors_skGen($\mathbf{SK}$.seed, $\mathbf{PK}$.seed, $\mathbf{ADRS}$, i)
3:      $\mathbf{ADRS}$.setTreeHeight(0)
4:      $\mathbf{ADRS}$.setTreeIndex(i)
5:      $node \leftarrow \mathbf{F}(\mathbf{PK}$.seed, $\mathbf{ADRS}$, $sk)$
6: **else**
7:      $lnode \leftarrow$ fors_node($\mathbf{SK}$.seed, $2i$, $z-1$, $\mathbf{PK}$.seed, $\mathbf{ADRS}$)
8:      $rnode \leftarrow$ fors_node($\mathbf{SK}$.seed, $2i+1$, $z-1$, $\mathbf{PK}$.seed, $\mathbf{ADRS}$)
9:      $\mathbf{ADRS}$.setTreeHeight(z)
10:      $\mathbf{ADRS}$.setTreeIndex(i)
11:      $node \leftarrow \mathbf{H}(\mathbf{PK}$.seed, $\mathbf{ADRS}$, $lnode \parallel rnode)$
12: **end if**
13: **return** $node$

8.3   Generating a FORS Signature

The fors_sign function (Algorithm 16) signs a $k \cdot a$-bit message digest md.[15] In addition to the message digest, fors_sign takes $\mathbf{SK}$.seed and $\mathbf{PK}$.seed from the SLH-DSA private key and an address as input. The address $\mathbf{ADRS}$ must have the layer address set to zero (since the XMSS

---

[15]For convenience, fors_sign takes a $\left\lceil \frac{k \cdot a}{8} \right\rceil$ byte message digest as input and extracts $k \cdot a$ bits to sign.

tree that signs a FORS key is always at layer 0), the tree address set to the XMSS tree that signs the FORS key, the *type* set to FORS_TREE, and the key pair address set to the index of the WOTS$^+$ key within the XMSS tree that signs the FORS key.

The fors_sign function splits $k \cdot a$ bits of md into k a-bit strings (line 2), each of which is interpreted as an integer between 0 and $t - 1$. Each of these integers is used to select a secret value from one of the k sets (line 4). For each secret value selected, an authentication path is computed and added to the signature (lines 5 through 9).

Algorithm 16 fors_sign(md, **SK**.seed, **PK**.seed, **ADRS**)

Generates a FORS signature.

Input: Message digest md, secret seed **SK**.seed, address **ADRS**, public seed **PK**.seed.
Output: FORS signature SIG$_{FORS}$.

1: SIG$_{FORS}$ = NULL $\triangleright$ initialize SIG$_{FORS}$ as a zero-length byte string
2: $indices \leftarrow$ base_2^b(md, a, k)
3: **for** i from 0 to $k - 1$ **do** $\triangleright$ compute signature elements
4:      SIG$_{FORS} \leftarrow$ SIG$_{FORS} \parallel$ fors_skGen(**SK**.seed, **PK**.seed, **ADRS**, $i \cdot 2^a + indices[i]$)

5:      **for** j from 0 to $a - 1$ **do** $\triangleright$ compute auth path
6:          $s \leftarrow \lfloor indices[i]/2^j \rfloor \oplus 1$
7:          AUTH$[j] \leftarrow$ fors_node(**SK**.seed, $i \cdot 2^{a-j} + s, j$, **PK**.seed, **ADRS**)
8:      **end for**
9:      SIG$_{FORS} \leftarrow$ SIG$_{FORS} \parallel$ AUTH
10: **end for**
11: **return** SIG$_{FORS}$

8.4 Computing a FORS Public Key From a Signature

Verifying a FORS signature involves computing a public-key value from a message digest and a signature value. Verification succeeds if the correct public-key value is computed, which is determined by verifying the hypertree signature on the computed public-key value using the SLH-DSA public key. This section describes fors_pkFromSig (Algorithm 17), a function that computes a candidate FORS public key from a FORS signature and corresponding message digest.

In addition to a message digest md and a $k \cdot (a + 1) \cdot n$-byte signature SIG$_{FORS}$, fors_pkFromSig takes **PK**.seed from the SLH-DSA public key and an address as input.[16] The address **ADRS** must have the layer address set to zero (since the XMSS tree that signs a FORS key is always at layer 0), the tree address set to the XMSS tree that signs the FORS key, the *type* set to FORS_TREE, and the key pair address set to the index of the WOTS$^+$ key within the XMSS tree that signs the FORS key.

The fors_pkFromSig function begins by computing the roots of each of the k Merkle trees (lines 2 through 20). As in fors_sign, $k \cdot a$ bits of the message digest are split into k a-bit strings (line 1), each of which is interpreted as an integer between 0 and $t - 1$. The integers are used to determine the locations in the Merkle trees of the secret values from the signature (lines 3

---

[16]As with fors_sign, fors_pkFromSig takes a $\lceil \frac{k \cdot a}{8} \rceil$ byte message digest as input and extracts $k \cdot a$ bits.

through 5). The hashes of the secret values are computed (line 6), and the hash values are used along with the corresponding authentication paths from the signature to compute the Merkle tree roots (lines 7 through 19). Once all of the Merkle tree roots have been computed, they are hashed together to compute the FORS public key (lines 21 through 24).

Algorithm 17 fors_pkFromSig(SIG$_{FORS}$, md, $\mathbf{PK}$.seed, $\mathbf{ADRS}$)

Computes a FORS public key from a FORS signature.

Input: FORS signature SIG$_{FORS}$, message digest md, public seed $\mathbf{PK}$.seed, address $\mathbf{ADRS}$.
Output: FORS public key.

1:   $indices \leftarrow \text{base\_}2^b(md, a, k)$
2:   **for** i from 0 to $k-1$ **do**
3:     $sk \leftarrow$ SIG$_{FORS}$.getSK(i) $\triangleright$ SIG$_{FORS}[i \cdot (a+1) \cdot n : (i \cdot (a+1)+1) \cdot n]$
4:     $\mathbf{ADRS}$.setTreeHeight(0) $\triangleright$ compute leaf
5:     $\mathbf{ADRS}$.setTreeIndex($i \cdot 2^a + indices[i]$)
6:     $node[0] \leftarrow \mathbf{F}(\mathbf{PK}$.seed, $\mathbf{ADRS}, sk)$

7:     $auth \leftarrow$ SIG$_{FORS}$.getAUTH(i) $\triangleright$ SIG$_{FORS}[(i \cdot (a+1)+1) \cdot n : (i+1) \cdot (a+1) \cdot n]$
8:     **for** j from 0 to $a-1$ **do** $\triangleright$ compute root from leaf and AUTH
9:       $\mathbf{ADRS}$.setTreeHeight($j+1$)
10:      **if** $\lfloor indices[i]/2^j \rfloor$ is even **then**
11:        $\mathbf{ADRS}$.setTreeIndex($\mathbf{ADRS}$.getTreeIndex()$/2$)
12:        $node[1] \leftarrow \mathbf{H}(\mathbf{PK}$.seed, $\mathbf{ADRS}, node[0] \parallel auth[j])$
13:      **else**
14:        $\mathbf{ADRS}$.setTreeIndex(($\mathbf{ADRS}$.getTreeIndex()$-1)/2$)
15:        $node[1] \leftarrow \mathbf{H}(\mathbf{PK}$.seed, $\mathbf{ADRS}, auth[j] \parallel node[0])$
16:      **end if**
17:      $node[0] \leftarrow node[1]$
18:     **end for**
19:     $root[i] \leftarrow node[0]$
20: **end for**
21: forspkADRS $\leftarrow \mathbf{ADRS}$ $\triangleright$ copy address to create a FORS public-key address
22: forspkADRS.setTypeAndClear(FORS_ROOTS)
23: forspkADRS.setKeyPairAddress($\mathbf{ADRS}$.getKeyPairAddress())
24: $pk \leftarrow \mathbf{T}_k(\mathbf{PK}$.seed, forspkADRS, $root)$ $\triangleright$ compute the FORS public key
25: **return** pk

---

9. SLH-DSA Internal Functions

SLH-DSA uses the hypertree and the FORS keys to create a stateless hash-based signature scheme. The SLH-DSA private key contains a secret seed value and a secret PRF key. The public key consists of a key identifier $PK.seed$ and the root of the hypertree. A signature is created by hashing the message, using part of the message digest to select a FORS key, signing other bits from the message digest with the FORS key, and generating a hypertree signature for the FORS key. The parameters for SLH-DSA are those specified previously for WOTS$^+$, XMSS, the SLH-DSA hypertree, and FORS (see Table 2).

SLH-DSA uses one additional parameter m, which is the length in bytes of the message digest. It is computed as:

$$m = \left\lceil \frac{h - h'}{8} \right\rceil + \left\lceil \frac{h'}{8} \right\rceil + \left\lceil \frac{k \cdot a}{8} \right\rceil$$

SLH-DSA uses h bits of the message digest to select a FORS key: $h - h'$ bits to select an XMSS tree at the lowest layer and h' bits to select a WOTS$^+$ key and corresponding FORS key from that tree. $k \cdot a$ bits of the digest are signed by the selected FORS key. While only $h + k \cdot a$ bits of the message digest are used, implementation is simplified by extracting the necessary bits from a slightly larger digest.

This section describes the functions for SLH-DSA key generation, signature generation, and signature verification. In the functions in this section, where randomness is required, the random values are provided as inputs to the functions. The interfaces specified in this section will be used when testing of SLH-DSA implementations is performed through the Cryptographic Algorithm Validation Program (CAVP). The key generation function in this section may also be used to obtain the assurance of private key possession via regeneration, as described in Section 3.1.

Other than for testing purposes, the interfaces for key generation and signature generation specified in this section **should not** be made available to applications, as any random values required for key generation and signature generation **shall** be generated by the cryptographic module. Section 10 provides guidance on the interfaces to be made available to applications.

9.1 SLH-DSA Key Generation

SLH-DSA public keys contain two elements (see Figure 16). The first is an n-byte public seed $PK.seed$, which is used in many hash function calls to provide domain separation between different SLH-DSA key pairs. The second value is the hypertree public key (i.e., the root of the top layer XMSS tree). $PK.seed$ **shall** be generated using an **approved** random bit generator (see

SK.seed	n bytes
SK.prf	n bytes
PK.seed	n bytes
PK.root	n bytes

Figure 15. SLH-DSA private key

PK.seed	n bytes
PK.root	n bytes

Figure 16. SLH-DSA public key

[14, 15, 16]), where the instantiation of the random bit generator supports at least $8n$ bits of security strength.

The SLH-DSA private key contains two random, secret n-byte values (see Figure 15). $\mathbf{SK}$.seed is used to generate all of the WOTS$^+$ and FORS private key elements. $\mathbf{SK}$.prf is used to generate a randomization value for the randomized hashing of the message in SLH-DSA. The private key also includes a copy of the public key. Both $\mathbf{SK}$.seed and $\mathbf{SK}$.prf **shall** be generated using an **approved** random bit generator, where the instantiation of the random bit generator supports at least $8n$ bits of security strength.

Algorithm 18 generates an SLH-DSA key pair. Lines 1 through 3 compute the root of the top layer XMSS tree. Line 4 bundles the three inputs and the computed $\mathbf{PK}$.seed into the private and public keys.

SLH-DSA signing has two variants — "hedged" and deterministic (see Section 9.2) — whose keys **should** only be used for the generation and verification of hedged and deterministic SLH-DSA digital signatures, respectively.

Algorithm 18 slh_keygen_internal($\mathbf{SK}$.seed, $\mathbf{SK}$.prf, $\mathbf{PK}$.seed)

Generates an SLH-DSA key pair.

Input: Secret seed $\mathbf{SK}$.seed, PRF key $\mathbf{SK}$.prf, public seed $\mathbf{PK}$.seed
Output: SLH-DSA key pair (SK, PK).

1: $\mathbf{ADRS} \leftarrow \mathrm{toByte}(0, 32)$ ▷ generate the public key for the top-level XMSS tree
2: $\mathbf{ADRS}.\mathrm{setLayerAddress}(d - 1)$
3: $\mathbf{PK}.\mathrm{root} \leftarrow \mathrm{xmss\_node}(\mathbf{SK}.\mathrm{seed}, 0, h', \mathbf{PK}.\mathrm{seed}, \mathbf{ADRS})$

4: **return** (($\mathbf{SK}$.seed, $\mathbf{SK}$.prf, $\mathbf{PK}$.seed, $\mathbf{PK}$.root), ($\mathbf{PK}$.seed, $\mathbf{PK}$.root))

9.2 SLH-DSA Signature Generation

An SLH-DSA signature consists of a randomization string, a FORS signature, and a hypertree signature, as shown in Figure 17.

Generating an SLH-DSA signature (Algorithm 19) begins by creating an m-byte message digest (lines 2 through 5). A PRF is used to create a message randomizer (line 3), and it is hashed along with the message to create the digest (line 5). Bits are then extracted from the message digest to be signed by the FORS key (line 6), to select an XMSS tree (lines 7 and 9), and to select a WOTS$^+$ key and corresponding FORS key within that XMSS tree (lines 8 and 10). Next, the FORS signature is computed (lines 11 through 14), and the corresponding FORS public key is obtained (line 16). Finally, the FORS public key is signed (line 17).

Randomness **R**	n bytes
FORS signature $\mathrm{SIG}_{\mathrm{FORS}}$	$k(1 + a) \cdot n$ bytes
HT signature $\mathrm{SIG}_{\mathrm{HT}}$	$(h + d \cdot len) \cdot n$ bytes

Figure 17. SLH-DSA signature data format

Algorithm 19 slh_sign_internal(M, SK, $addrnd$)

Generates an SLH-DSA signature.

Input: Message M, private key SK = (**SK**.seed, **SK**.prf, **PK**.seed, **PK**.root),
    (optional) additional randomness $addrnd$.
Output: SLH-DSA signature SIG.

1: $\mathbf{ADRS} \leftarrow \text{toByte}(0, 32)$

2: $opt\_rand \leftarrow addrnd$ $\quad \triangleright$ substitute $opt\_rand \leftarrow$ **PK**.seed for the deterministic variant
3: $R \leftarrow \mathbf{PRF}_{msg}(\mathbf{SK}.prf, opt\_rand, M)$ $\qquad\qquad\qquad\qquad \triangleright$ generate randomizer
4: $\text{SIG} \leftarrow R$

5: $digest \leftarrow \mathbf{H}_{msg}(R, \mathbf{PK}.seed, \mathbf{PK}.root, M)$ $\qquad\qquad \triangleright$ compute message digest
6: $md \leftarrow digest\left[0 : \left\lceil \frac{k \cdot a}{8} \right\rceil\right]$ $\qquad\qquad\qquad\qquad\qquad \triangleright$ first $\left\lceil \frac{k \cdot a}{8} \right\rceil$ bytes
7: $tmp\_idx_{tree} \leftarrow digest\left[\left\lceil \frac{k \cdot a}{8} \right\rceil : \left\lceil \frac{k \cdot a}{8} \right\rceil + \left\lceil \frac{h-h/d}{8} \right\rceil\right]$ $\qquad \triangleright$ next $\left\lceil \frac{h-h/d}{8} \right\rceil$ bytes
8: $tmp\_idx_{leaf} \leftarrow digest\left[\left\lceil \frac{k \cdot a}{8} \right\rceil + \left\lceil \frac{h-h/d}{8} \right\rceil : \left\lceil \frac{k \cdot a}{8} \right\rceil + \left\lceil \frac{h-h/d}{8} \right\rceil + \left\lceil \frac{h}{8d} \right\rceil\right]$ $\triangleright$ next $\left\lceil \frac{h}{8d} \right\rceil$ bytes

9: $idx_{tree} \leftarrow \text{toInt}\left(tmp\_idx_{tree}, \left\lceil \frac{h-h/d}{8} \right\rceil\right) \bmod 2^{h-h/d}$
10: $idx_{leaf} \leftarrow \text{toInt}\left(tmp\_idx_{leaf}, \left\lceil \frac{h}{8d} \right\rceil\right) \bmod 2^{h/d}$

11: $\mathbf{ADRS}.\text{setTreeAddress}(idx_{tree})$
12: $\mathbf{ADRS}.\text{setTypeAndClear}(\text{FORS\_TREE})$
13: $\mathbf{ADRS}.\text{setKeyPairAddress}(idx_{leaf})$
14: $\text{SIG}_{FORS} \leftarrow \text{fors\_sign}(md, \mathbf{SK}.seed, \mathbf{PK}.seed, \mathbf{ADRS})$
15: $\text{SIG} \leftarrow \text{SIG} \parallel \text{SIG}_{FORS}$

16: $\text{PK}_{FORS} \leftarrow \text{fors\_pkFromSig}(\text{SIG}_{FORS}, md, \mathbf{PK}.seed, \mathbf{ADRS})$ $\qquad\qquad \triangleright$ get FORS key

17: $\text{SIG}_{HT} \leftarrow \text{ht\_sign}(\text{PK}_{FORS}, \mathbf{SK}.seed, \mathbf{PK}.seed, idx_{tree}, idx_{leaf})$
18: $\text{SIG} \leftarrow \text{SIG} \parallel \text{SIG}_{HT}$
19: **return** SIG

The message randomizer may be set in either a non-deterministic or deterministic way, depending on whether $addrnd$ is provided as an input. For the "hedged" variant, $addrnd$ is provided as an input, and opt_rand is set to $addrnd$. The hedged variant is the default and **should** be used on platforms where side-channel attacks are a concern. When using the hedged version, $addrnd$ **shall** be an n-byte random value. While $addrnd$ **should** ideally be generated by an **approved** random bit generator, other methods for generating fresh random values may be used. For the deterministic variant, $addrnd$ is not provided as an input, and opt_rand is set to $\mathbf{PK}.seed$, which results in signing being deterministic (i.e., signing the same message twice will result in the same signature). The deterministic variant is available for platforms where a random bit generator is not available.

9.3 SLH-DSA Signature Verification

As with signature generation, SLH-DSA signature verification (Algorithm 20) begins by computing a message digest (line 8) and then extracting md (line 9), idx_{tree} (lines 10 and 12), and idx_{leaf} (lines 11 and 13) from the digest. A candidate FORS public key is then computed (line 17), and the signature on the FORS key is verified (line 18). If this signature verification succeeds, then the correct FORS public key was computed, and the signature SIG on message M is valid.

Algorithm 20 slh_verify_internal(M, SIG, PK)

Verifies an SLH-DSA signature.

Input: Message M, signature SIG, public key PK = ($\mathbf{PK}.seed$, $\mathbf{PK}.root$).
Output: Boolean.

1: **if** $|\text{SIG}| \neq (1 + k(1 + a) + h + d \cdot len) \cdot n$ **then**
2:      **return** false
3: **end if**
4: $\mathbf{ADRS} \leftarrow$ toByte$(0, 32)$
5: $R \leftarrow$ SIG.getR() $\triangleright$ SIG$[0 : n]$
6: SIG$_{FORS} \leftarrow$ SIG.getSIG_FORS() $\triangleright$ SIG$[n : (1 + k(1 + a)) \cdot n]$
7: SIG$_{HT} \leftarrow$ SIG.getSIG_HT() $\triangleright$ SIG$[(1 + k(1 + a)) \cdot n : (1 + k(1 + a) + h + d \cdot len) \cdot n]$
8: $digest \leftarrow \mathbf{H}_{msg}(R, \mathbf{PK}.seed, \mathbf{PK}.root, M)$ $\triangleright$ compute message digest
9: $md \leftarrow digest\left[0 : \left\lceil\frac{k \cdot a}{8}\right\rceil\right]$ $\triangleright$ first $\left\lceil\frac{k \cdot a}{8}\right\rceil$ bytes
10: $tmp\_idx_{tree} \leftarrow digest\left[\left\lceil\frac{k \cdot a}{8}\right\rceil : \left\lceil\frac{k \cdot a}{8}\right\rceil + \left\lceil\frac{h - h/d}{8}\right\rceil\right]$ $\triangleright$ next $\left\lceil\frac{h - h/d}{8}\right\rceil$ bytes
11: $tmp\_idx_{leaf} \leftarrow digest\left[\left\lceil\frac{k \cdot a}{8}\right\rceil + \left\lceil\frac{h - h/d}{8}\right\rceil : \left\lceil\frac{k \cdot a}{8}\right\rceil + \left\lceil\frac{h - h/d}{8}\right\rceil + \left\lceil\frac{h}{8d}\right\rceil\right]$ $\triangleright$ next $\left\lceil\frac{h}{8d}\right\rceil$ bytes
12: $idx_{tree} \leftarrow$ toInt$\left(tmp\_idx_{tree}, \left\lceil\frac{h - h/d}{8}\right\rceil\right)$ mod $2^{h-h/d}$
13: $idx_{leaf} \leftarrow$ toInt$\left(tmp\_idx_{leaf}, \left\lceil\frac{h}{8d}\right\rceil\right)$ mod $2^{h/d}$
14: $\mathbf{ADRS}$.setTreeAddress(idx_{tree}) $\triangleright$ compute FORS public key
15: $\mathbf{ADRS}$.setTypeAndClear(FORS_TREE)
16: $\mathbf{ADRS}$.setKeyPairAddress(idx_{leaf})
17: PK$_{FORS} \leftarrow$ fors_pkFromSig(SIG$_{FORS}$, md, $\mathbf{PK}.seed$, $\mathbf{ADRS}$)
18: **return** ht_verify(PK$_{FORS}$, SIG$_{HT}$, $\mathbf{PK}.seed$, idx_{tree}, idx_{leaf}, $\mathbf{PK}.root$)

---

10. SLH-DSA External Functions

This section provides guidance on the key generation, signature generation, and signature verification functions that should be provided for use by applications. The functions in this section use the functions in Section 9 to implement the underlying SLH-DSA scheme.

10.1 SLH-DSA Key Generation

Algorithm 21 generates an SLH-DSA key pair. Lines 1 through 3 generate the random values for the private and public keys, and line 7 calls slh_keygen_internal to compute $PK.root$ and return the private and public key. $PK.seed$, $SK.seed$, and $SK.prf$ **shall** be generated using an **approved** random bit generator (see [14, 15, 16]), where the instantiation of the random bit generator supports at least $8n$ bits of security strength.

Algorithm 21 slh_keygen()

Generates an SLH-DSA key pair.

Input: (none)
Output: SLH-DSA key pair (SK, PK).

1: $SK.seed \xleftarrow{\$} \mathbb{B}^n$ ▷ set $SK.seed$, $SK.prf$, and $PK.seed$ to random n-byte
2: $SK.prf \xleftarrow{\$} \mathbb{B}^n$ ▷ strings using an **approved** random bit generator
3: $PK.seed \xleftarrow{\$} \mathbb{B}^n$

4: **if** $SK.seed = $ NULL **or** $SK.prf = $ NULL **or** $PK.seed = $ NULL **then**
5:     **return** $\perp$ ▷ return an error indication if random bit generation failed
6: **end if**

7: **return** slh_keygen_internal($SK.seed$, $SK.prf$, $PK.seed$)

10.2 SLH-DSA Signature Generation

This section presents two versions of SLH-DSA signature generation: a "pure" version (slh_sign) and a "pre-hash" version (hash_slh_sign). Both versions use slh_sign_internal, but they differ in how the message input to slh_sign_internal is created from the content to be signed. In the pure version, the content is signed by slh_sign_internal along with some domain separation information. In the pre-hash version, a hash of the content is signed by slh_sign_internal along with some domain separation information.

Both versions take the content to be signed, the private key, and a context as input. The pre-hash version also takes as input a hash function or XOF that is to be used to pre-hash the content to be signed. The context string has a maximum length of 255 bytes. By default, the context is the empty string. However, applications may specify the use of a non-empty context string.

The identifier for a signature (e.g., the object identifier [OID]) **should** indicate whether the signature is a pure signature or a pre-hash signature. In the case of pre-hash signatures, the identifier **should** also indicate the hash function or XOF used to compute the pre-hash.[17] While

---

[17]In the case of a XOF, this would also include the length of the output from the XOF.

a single key pair may be used for both pure and pre-hash signatures, it is recommended that each key pair only be used for one version or the other. If a non-empty context string is to be used, this should either be indicated by the signature's identifier or the application with which the signature is being used.

If the default hedged variant of slh_sign_internal is used, the n-byte random value $addrnd$ **shall** be generated by the cryptographic module that runs slh_sign_internal. However, M' in Algorithms 22 and 23 may be constructed outside of the crytographic module. In the case of hash_slh_sign, the hash or XOF of the content to be signed must be computed within a FIPS 140-validated cryptographic module, but it may be a different cryptographic module than the one that runs slh_sign_internal.

In general, the pure version is preferred. However, for some cryptographic modules that generate SLH-DSA signatures, performing lines 3 and 5 of Algorithm 19 may be infeasible if the message M is large. This may, for example, be the result of the module having limited memory to store the message to be signed. Similarly, for some cryptographic modules that verify SLH-DSA signatures, performing line 8 of Algorithm 20 may be infeasible if the message M is large. For some use cases, these issues may be addressed by signing a digest of the content rather than signing the content directly.

In many cases where the content to be signed is large, hashing of the content is performed at the application level. For example, in the Cryptographic Message Syntax [23], a digest of the content may be computed, and that digest is signed along with other attributes. In cases in which the content is not hashed at the application level, the pre-hash version of SLH-DSA signing (Section 10.2.2) may be used.

To maintain the same level of security strength when the content is hashed at the application level or when using the pre-hash version of SLH-DSA, the digest that is signed needs to be generated using an **approved** hash function or XOF (e.g., from FIPS 180-4 [8] or FIPS 202 [6]) that provides at least $8n$ bits of classical security strength against both collision and second preimage attacks [6, Table 4].[18] Verification of a signature created in this way will require the verify function to generate a digest from the message in the same way for input to the verification function.

Even if it is feasible to compute collisions on the hash functions or XOF used to instantiate $\mathbf{H}_{msg}$, $\mathbf{PRF}$, $\mathbf{PRF}_{msg}$, $\mathbf{F}$, $\mathbf{H}$, and $\mathbf{T}_l$, there is believed to be no adverse effect on the security of SLH-DSA.[19] However, if the input to the signing function is a digest of the content, then collisions on the function used to compute the digest can result in forged messages.

10.2.1 Pure SLH-DSA Signature Generation

In the pure version, the content to be signed is prepended with a one-byte domain separator, one byte that indicates the length of the context string, and the context string. The domain separator, which has a value of zero for pure signing, is included to prevent pre-hash signatures from verifying as pure signatures and vice versa. In the default case in which the context string

---

[18]Obtaining at least $8n$ bits of classical security strength against collision attacks requires that the digest to be signed is at least $2n$ bytes in length.

[19]As noted in Section 11, applications that require message-bound signatures may be adversely affected if it is feasible to compute collisions on $\mathbf{H}_{msg}$.

is empty, pure signing simply involves prepending two zero bytes to the content to be signed and passing the result to slh_sign_internal along with the private key and, in the case of hedged signing, an n-byte random value.

Algorithm 22 slh_sign(M, ctx, SK)

Generates a pure SLH-DSA signature.

Input: Message M, context string ctx, private key SK.
Output: SLH-DSA signature SIG.

1: **if** $|ctx| > 255$ **then**
2:     **return** $\perp$ ▷ return an error indication if the context string is too long
3: **end if**

4: $addrnd \xleftarrow{\$} \mathbb{B}^n$ ▷ skip lines 4 through 7 for the deterministic variant
5: **if** $addrnd =$ NULL **then**
6:     **return** $\perp$ ▷ return an error indication if random bit generation failed
7: **end if**

8: $M' \leftarrow \text{toByte}(0,1) \parallel \text{toByte}(|ctx|,1) \parallel ctx \parallel M$
9: SIG $\leftarrow$ slh_sign_internal(M', SK, $addrnd$)   ▷ omit $addrnd$ for the deterministic variant
10: **return** SIG

10.2.2 HashSLH-DSA Signature Generation

In the pre-hash version, the message input to slh_sign_internal is the result of applying either a hash function or a XOF to the content to be signed. The output of the hash function or XOF is prepended by a one-byte domain separator, one byte that indicates the length of the context string, the context string, and the distinguished encoding rules (DER) encoding of the hash function or XOF's OID. The domain separator has a value of one for pre-hash signing. The DER encoding of the OID includes the tag and length.

Algorithm 23 shows the DER encodings of the OIDs for SHA-256, SHA-512, SHAKE128, and SHAKE256. However, hash_slh_sign may be used with other hash functions or XOFs. SHA-256 and SHAKE128 are only appropriate for use with SLH-DSA parameter sets that are claimed to be in security category 1 (see Section 11).

Algorithm 23 hash_slh_sign(M, ctx, PH, SK)

Generates a pre-hash SLH-DSA signature.

Input: Message M, context string ctx, pre-hash function PH, private key SK.
Output: SLH-DSA signature SIG.

1: **if** $|ctx| > 255$ **then**
2:      **return** $\perp$ $\triangleright$ return an error indication if the context string is too long
3: **end if**

4: $addrnd \xleftarrow{\$} \mathbb{B}^n$ $\triangleright$ skip lines 4 through 7 for the deterministic variant
5: **if** $addrnd =$ NULL **then**
6:      **return** $\perp$ $\triangleright$ return an error indication if random bit generation failed
7: **end if**

8: **switch** PH **do**
9:      **case** SHA-256:
10:          OID $\leftarrow$ toByte(0x0609608648016503040201, 11) $\triangleright$ 2.16.840.1.101.3.4.2.1
11:          $\text{PH}_M \leftarrow$ SHA-256(M)
12:      **case** SHA-512:
13:          OID $\leftarrow$ toByte(0x0609608648016503040203, 11) $\triangleright$ 2.16.840.1.101.3.4.2.3
14:          $\text{PH}_M \leftarrow$ SHA-512(M)
15:      **case** SHAKE128:
16:          OID $\leftarrow$ toByte(0x060960864801650304020B, 11) $\triangleright$ 2.16.840.1.101.3.4.2.11
17:          $\text{PH}_M \leftarrow$ SHAKE128(M, 256)
18:      **case** SHAKE256:
19:          OID $\leftarrow$ toByte(0x060960864801650304020C, 11) $\triangleright$ 2.16.840.1.101.3.4.2.12
20:          $\text{PH}_M \leftarrow$ SHAKE256(M, 512)
21:      **case** ... $\triangleright$ other **approved** hash functions or XOFs
22:          ...
23: **end switch**
24: $M' \leftarrow$ toByte(1, 1) $\|$ toByte($|ctx|$, 1) $\|$ ctx $\|$ OID $\|$ PH_M
25: SIG $\leftarrow$ slh_sign_internal(M', SK, $addrnd$) $\triangleright$ omit $addrnd$ for the deterministic variant
26: **return** SIG

10.3 SLH-DSA Signature Verification

Algorithms 24 and 25 present the pure and pre-hash versions of SLH-DSA signature verification that correspond to the pure and pre-hash versions of SLH-DSA signature generation in Section 10.2. These functions construct M' in the same way as Algorithms 22 and 23, respectively, and pass the resulting M' to slh_verify_internal for verification. As with signature generation, M' may be constructed outside of the cryptographic module that performs slh_verify_internal. However, in the case of hash_slh_verify, the hash or XOF of the content must be computed within a FIPS 140-validated cryptographic module, which may be a different cryptographic module than the one that performs slh_verify_internal.

The identifier associated with the signature should indicate whether the pure or pre-hash version of signature verification should be used, and in the pre-hash case, the hash function or XOF to use to compute the pre-hash. A non-empty context string should be used in verification if one is specified in the signature's identifier or by the application with which the signature is being used.

Algorithm 24 slh_verify(M, SIG, ctx, PK)

Verifies a pure SLH-DSA signature.

Input: Message M, signature SIG, context string ctx, public key PK.
Output: Boolean.

1: **if** $|ctx| > 255$ **then**
2:     **return** false
3: **end if**

4: $M' \leftarrow \text{toByte}(0,1) \parallel \text{toByte}(|ctx|,1) \parallel ctx \parallel M$
5: **return** slh_verify_internal(M', SIG, PK)

Algorithm 25 hash_slh_verify(M, SIG, ctx, PH, PK)

Verifies a pre-hash SLH-DSA signature.

Input: Message M, signature SIG, context string ctx, pre-hash function PH, public key PK.
Output: Boolean.

1: **if** $|ctx| > 255$ **then**
2:    **return** false
3: **end if**

4: **switch** PH **do**
5:    **case** SHA-256:
6:        OID $\leftarrow$ toByte($0\mathrm{x}0609608648016503040201, 11$) $\triangleright$ 2.16.840.1.101.3.4.2.1
7:        $\text{PH}_M \leftarrow$ SHA-256(M)

8:    **case** SHA-512:
9:        OID $\leftarrow$ toByte($0\mathrm{x}0609608648016503040203, 11$) $\triangleright$ 2.16.840.1.101.3.4.2.3
10:        $\text{PH}_M \leftarrow$ SHA-512(M)

11:    **case** SHAKE128:
12:        OID $\leftarrow$ toByte($0\mathrm{x}060960864801650304020\mathrm{B}, 11$) $\triangleright$ 2.16.840.1.101.3.4.2.11
13:        $\text{PH}_M \leftarrow$ SHAKE128($M, 256$)

14:    **case** SHAKE256:
15:        OID $\leftarrow$ toByte($0\mathrm{x}060960864801650304020\mathrm{C}, 11$) $\triangleright$ 2.16.840.1.101.3.4.2.12
16:        $\text{PH}_M \leftarrow$ SHAKE256($M, 512$)

17:    **case** ... $\triangleright$ other **approved** hash functions or XOFs
18:        ...
19: **end switch**
20: $M' \leftarrow$ toByte($1, 1$) || toByte($|ctx|, 1$) || ctx || OID || PH_M
21: **return** slh_verify_internal(M', SIG, PK)

42

11. Parameter Sets

This standard approves 12 parameter sets for use with SLH-DSA. A parameter set consists of parameters for WOTS$^+$ (n and lg_w), XMSS and the SLH-DSA hypertree (h and d), and FORS (k and a), as well as instantiations for the functions H_{msg}, PRF, PRF_{msg}, F, H, and T_ℓ.

Table 2 lists the parameter sets that are **approved** for use.[20] Each parameter set name indicates the hash function family (SHA2 or SHAKE) that is used to instantiate the hash functions, the length in bits of the security parameter n, and whether the parameter set was designed to create relatively small signatures ('s') or to have relatively fast signature generation ('f'). There are six sets of values for n, lg_w, h, d, k, and a that are **approved** for use.[21] For each of the six sets of values, the functions H_{msg}, PRF, PRF_{msg}, F, H, and T_ℓ may be instantiated using either SHAKE [6] or SHA-2 [8]. For the SHAKE parameter sets, the functions **shall** be instantiated as specified in Section 11.1. For the SHA2 parameter sets, the functions **shall** be instantiated as specified in Section 11.2.1 if $n = 16$ and **shall** be instantiated as specified in Section 11.2.2 if $n = 24$ or $n = 32$.

Table 2. SLH-DSA parameter sets

	n	h	d	h'	a	k	lg_w	m	security category	pk bytes	sig bytes
SLH-DSA-SHA2-128s SLH-DSA-SHAKE-128s	16	63	7	9	12	14	4	30	1	32	7 856
SLH-DSA-SHA2-128f SLH-DSA-SHAKE-128f	16	66	22	3	6	33	4	34	1	32	17 088
SLH-DSA-SHA2-192s SLH-DSA-SHAKE-192s	24	63	7	9	14	17	4	39	3	48	16 224
SLH-DSA-SHA2-192f SLH-DSA-SHAKE-192f	24	66	22	3	8	33	4	42	3	48	35 664
SLH-DSA-SHA2-256s SLH-DSA-SHAKE-256s	32	64	8	8	14	22	4	47	5	64	29 792
SLH-DSA-SHA2-256f SLH-DSA-SHAKE-256f	32	68	17	4	9	35	4	49	5	64	49 856

The 12 parameter sets included in Table 2 were designed to meet certain security strength categories defined by NIST in its original Call for Proposals [25] with respect to existential unforgeability under chosen message attack (EUF-CMA) when each key pair is used to sign at most 2^{64} messages.[22] These security strength categories are explained further in SP 800-57, Part 1 [9].

---

[20]SP 800-230 [24] specifies additional parameter sets that are **approved** for use. While key pairs generated for the parameter sets specified in this standard may be used to sign up to 2^{64} messages, key pairs generated for the parameter sets in SP 800-230 are more limited in the number of signatures that may be generated.

[21]In addition to n, lg_w, h, d, k, and a, Table 2 also lists values for parameters that may be computed from these values (h', m, public-key size(pk bytes), and signature size(sig bytes)). The security category is the security category in which the parameter set is claimed to be [10].

[22]If a key pair were used to sign 10 billion (10^{10}) messages per second, it would take over 58 years to sign 2^{64} messages.

Using this approach, security strength is not described by a single number, such as "128 bits of security." Instead, each parameter set is claimed to be at least as secure as a generic block cipher with a prescribed key size. More precisely, it is claimed that the computational resources needed to break SLH-DSA are greater than or equal to the computational resources needed to break the block cipher when these computational resources are estimated using any realistic model of computation. Different models of computation can be more or less realistic and, accordingly, lead to more or less accurate estimates of security strength. Some commonly studied models are discussed in [26].

Concretely, the parameter sets with $n = 16$ are claimed to be in security category 1, the parameter sets with $n = 24$ are claimed to be in security category 3, and the parameter sets with $n = 32$ are claimed to be in security category 5 [10]. For additional discussion of the security strength of SLH-DSA, see [10, 27].

Some applications require a property known as message-bound signatures [28, 29], which intuitively requires that it be infeasible for anyone to create a public key and a signature that are valid for two different messages. Signature schemes are not required to have this property under the EUF-CMA security definition used in assigning security categories. In the case of SLH-DSA, the key pair owner could create two messages with the same signature by finding a collision on $\mathbf{H}_{msg}$. Due to the length of the output of $\mathbf{H}_{msg}$, finding such a collision would be expected to require fewer computational resources than specified for the parameter sets' claimed security categories in all cases except SLH-DSA-SHA2-128f and SLH-DSA-SHAKE-128f.[23] Therefore, applications that require message-bound signatures should either take the expected cost of finding collisions on $\mathbf{H}_{msg}$ into account when choosing an appropriate parameter set or apply a technique (e.g., the BUFF transformation [29]) to obtain the message-bound signatures property.

11.1 SLH-DSA Using SHAKE

$\mathbf{H}_{msg}$, $\mathbf{PRF}$, $\mathbf{PRF}_{msg}$, $\mathbf{F}$, $\mathbf{H}$, and $\mathbf{T}_\ell$ **shall** be instantiated as follows for the SLH-DSA-SHAKE-128s, SLH-DSA-SHAKE-128f, SLH-DSA-SHAKE-192s, SLH-DSA-SHAKE-192f, SLH-DSA-SHAKE-256s, and SLH-DSA-SHAKE-256f parameter sets:

$\mathbf{H}_{msg}(R, \mathbf{PK.seed}, \mathbf{PK.root}, M) = \mathsf{SHAKE256}(R \parallel \mathbf{PK.seed} \parallel \mathbf{PK.root} \parallel M, 8m)$
$\mathbf{PRF}(\mathbf{PK.seed}, \mathbf{SK.seed}, \mathbf{ADRS}) = \mathsf{SHAKE256}(\mathbf{PK.seed} \parallel \mathbf{ADRS} \parallel \mathbf{SK.seed}, 8n)$
$\mathbf{PRF}_{msg}(\mathbf{SK.prf}, opt\_rand, M) = \mathsf{SHAKE256}(\mathbf{SK.prf} \parallel opt\_rand \parallel M, 8n)$
$\mathbf{F}(\mathbf{PK.seed}, \mathbf{ADRS}, M_1) = \mathsf{SHAKE256}(\mathbf{PK.seed} \parallel \mathbf{ADRS} \parallel M_1, 8n)$
$\mathbf{H}(\mathbf{PK.seed}, \mathbf{ADRS}, M_2) = \mathsf{SHAKE256}(\mathbf{PK.seed} \parallel \mathbf{ADRS} \parallel M_2, 8n)$
$\mathbf{T}_\ell(\mathbf{PK.seed}, \mathbf{ADRS}, M_\ell) = \mathsf{SHAKE256}(\mathbf{PK.seed} \parallel \mathbf{ADRS} \parallel M_\ell, 8n)$

11.2 SLH-DSA Using SHA2

In Sections 11.2.1 and 11.2.2, the functions MGF1-SHA-256 and MGF1-SHA-512 are MGF1 from Appendix B.2.1 of RFC 8017 [30], where Hash is SHA-256 or SHA-512, respectively. The functions

---

[23]Finding a collision would be expected to require computing $\mathbf{H}_{msg}$ for approximately $2^{(h+k \cdot a)/2}$ different messages.

HMAC-SHA-256 and HMAC-SHA-512 are the HMAC function from FIPS 198-1 [31, 32], where H is SHA-256 or SHA-512, respectively.

The functions in Sections 11.2.1 and 11.2.2 make use of a compressed version of $\mathbf{ADRS}$ (see Figure 18). A compressed address ($\mathbf{ADRS}^c$) is a 22-byte string that is the same as an $\mathbf{ADRS}$ with the exceptions that the encodings of the layer address and type are reduced to one byte each and the encoding of the tree address is reduced to eight bytes (i.e., $\mathbf{ADRS}^c = \mathbf{ADRS}[3] \parallel \mathbf{ADRS}[8:16] \parallel \mathbf{ADRS}[19] \parallel \mathbf{ADRS}[20:32]$). For implementations of the SHA2 parameter sets that store addresses in compressed form (i.e., 22 bytes), the member functions (Section 4.3) are as shown in Table 3 rather than Table 1.

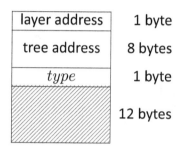

Figure 18. Compressed address ($\mathbf{ADRS}^c$)

Table 3. Member functions for compressed addresses

Member function	Expanded notation
$\mathbf{ADRS}$.setLayerAddress(l)	$\mathbf{ADRS} \leftarrow \text{toByte}(l, 1) \parallel \mathbf{ADRS}[1:22]$
$\mathbf{ADRS}$.setTreeAddress(t)	$\mathbf{ADRS} \leftarrow \mathbf{ADRS}[0:1] \parallel \text{toByte}(t, 8) \parallel \mathbf{ADRS}[9:22]$
$\mathbf{ADRS}$.setTypeAndClear(Y)	$\mathbf{ADRS} \leftarrow \mathbf{ADRS}[0:9] \parallel \text{toByte}(Y, 1) \parallel \text{toByte}(0, 12)$
$\mathbf{ADRS}$.setKeyPairAddress(i)	$\mathbf{ADRS} \leftarrow \mathbf{ADRS}[0:10] \parallel \text{toByte}(i, 4) \parallel \mathbf{ADRS}[14:22]$
$\mathbf{ADRS}$.setChainAddress(i) $\mathbf{ADRS}$.setTreeHeight(i)	$\mathbf{ADRS} \leftarrow \mathbf{ADRS}[0:14] \parallel \text{toByte}(i, 4) \parallel \mathbf{ADRS}[18:22]$
$\mathbf{ADRS}$.setHashAddress(i) $\mathbf{ADRS}$.setTreeIndex(i)	$\mathbf{ADRS} \leftarrow \mathbf{ADRS}[0:18] \parallel \text{toByte}(i, 4)$
$i \leftarrow \mathbf{ADRS}$.getKeyPairAddress()	$i \leftarrow \text{toInt}(\mathbf{ADRS}[10:14], 4)$
$i \leftarrow \mathbf{ADRS}$.getTreeIndex()	$i \leftarrow \text{toInt}(\mathbf{ADRS}[18:22], 4)$

11.2.1 SLH-DSA Using SHA2 for Security Category 1

$\mathbf{H}_{msg}$, $\mathbf{PRF}$, $\mathbf{PRF}_{msg}$, $\mathbf{F}$, $\mathbf{H}$, and $\mathbf{T}_\ell$ **shall** be instantiated as follows for the SLH-DSA-SHA2-128s and SLH-DSA-SHA2-128f parameter sets:

$$\mathbf{H}_{msg}(R, \mathbf{PK}.\text{seed}, \mathbf{PK}.\text{root}, M) = $$
$$\text{MGF1-SHA-256}(R \parallel \mathbf{PK}.\text{seed} \parallel \text{SHA-256}(R \parallel \mathbf{PK}.\text{seed} \parallel \mathbf{PK}.\text{root} \parallel M), m)$$

$\mathrm{PRF}(\mathbf{PK.seed}, \mathbf{SK.seed}, \mathbf{ADRS}) =$
$\qquad \mathsf{Trunc}_n(\text{SHA-256}(\mathbf{PK.seed} \parallel \mathsf{toByte}(0, 64 - n) \parallel \mathbf{ADRS}^c \parallel \mathbf{SK.seed}))$

$\mathrm{PRF}_{msg}(\mathbf{SK.prf}, opt\_rand, M) =$
$\qquad \mathsf{Trunc}_n(\text{HMAC-SHA-256}(\mathbf{SK.prf}, opt\_rand \parallel M))$

$\mathrm{F}(\mathbf{PK.seed}, \mathbf{ADRS}, M_1) =$
$\qquad \mathsf{Trunc}_n(\text{SHA-256}(\mathbf{PK.seed} \parallel \mathsf{toByte}(0, 64 - n) \parallel \mathbf{ADRS}^c \parallel M_1))$

$\mathrm{H}(\mathbf{PK.seed}, \mathbf{ADRS}, M_2) =$
$\qquad \mathsf{Trunc}_n(\text{SHA-256}(\mathbf{PK.seed} \parallel \mathsf{toByte}(0, 64 - n) \parallel \mathbf{ADRS}^c \parallel M_2))$

$\mathrm{T}_\ell(\mathbf{PK.seed}, \mathbf{ADRS}, M_\ell) =$
$\qquad \mathsf{Trunc}_n(\text{SHA-256}(\mathbf{PK.seed} \parallel \mathsf{toByte}(0, 64 - n) \parallel \mathbf{ADRS}^c \parallel M_\ell))$

11.2.2 SLH-DSA Using SHA2 for Security Categories 3 and 5

$\mathrm{H}_{msg}, \mathrm{PRF}, \mathrm{PRF}_{msg}, \mathrm{F}, \mathrm{H}$, and T_ℓ **shall** be instantiated as follows for the SLH-DSA-SHA2-192s, SLH-DSA-SHA2-192f, SLH-DSA-SHA2-256s, and SLH-DSA-SHA2-256f parameter sets:

$\mathrm{H}_{msg}(R, \mathbf{PK.seed}, \mathbf{PK.root}, M) =$
$\qquad \text{MGF1-SHA-512}(R \parallel \mathbf{PK.seed} \parallel \text{SHA-512}(R \parallel \mathbf{PK.seed} \parallel \mathbf{PK.root} \parallel M), m)$

$\mathrm{PRF}(\mathbf{PK.seed}, \mathbf{SK.seed}, \mathbf{ADRS}) =$
$\qquad \mathsf{Trunc}_n(\text{SHA-256}(\mathbf{PK.seed} \parallel \mathsf{toByte}(0, 64 - n) \parallel \mathbf{ADRS}^c \parallel \mathbf{SK.seed}))$

$\mathrm{PRF}_{msg}(\mathbf{SK.prf}, opt\_rand, M) =$
$\qquad \mathsf{Trunc}_n(\text{HMAC-SHA-512}(\mathbf{SK.prf}, opt\_rand \parallel M))$

$\mathrm{F}(\mathbf{PK.seed}, \mathbf{ADRS}, M_1) =$
$\qquad \mathsf{Trunc}_n(\text{SHA-256}(\mathbf{PK.seed} \parallel \mathsf{toByte}(0, 64 - n) \parallel \mathbf{ADRS}^c \parallel M_1))$

$\mathrm{H}(\mathbf{PK.seed}, \mathbf{ADRS}, M_2) =$
$\qquad \mathsf{Trunc}_n(\text{SHA-512}(\mathbf{PK.seed} \parallel \mathsf{toByte}(0, 128 - n) \parallel \mathbf{ADRS}^c \parallel M_2))$

$\mathrm{T}_\ell(\mathbf{PK.seed}, \mathbf{ADRS}, M_\ell) =$
$\qquad \mathsf{Trunc}_n(\text{SHA-512}(\mathbf{PK.seed} \parallel \mathsf{toByte}(0, 128 - n) \parallel \mathbf{ADRS}^c \parallel M_\ell))$

References

[1] National Institute of Standards and Technology (2023) Digital Signature Standard (DSS). (Department of Commerce, Washington, D.C.), Federal Information Processing Standards Publications (FIPS) NIST FIPS 186-5. https://doi.org/10.6028/NIST.FIPS.186-5.

[2] National Institute of Standards and Technology (2024) Guideline for Using Cryptographic Standards in the Federal Government: Cryptographic Mechanisms. (National Institute of Standards and Technology, Gaithersburg, MD), NIST Special Publication (SP) 800-175B, Rev. 2. [Forthcoming: will be available at https://csrc.nist.gov/publications].

[3] National Institute of Standards and Technology (2024) Recommendation for Obtaining Assurances for Digital Signature Applications. (National Institute of Standards and Technology, Gaithersburg, MD), NIST Special Publication (SP) 800-89, Rev. 1. [Forthcoming: will be available at https://csrc.nist.gov/publications].

[4] Aumasson JP, Bernstein DJ, Beullens W, Dobraunig C, Eichlseder M, Fluhrer S, Gazdag SL, Hülsing A, Kampanakis P, Kölbl S, Lange T, Lauridsen MM, Mendel F, Niederhagen R, Rechberger C, Rijneveld J, Schwabe P, Westerbaan B (2020) SPHINCS$^+$ – Submission to the NIST post-quantum project, v.3.

[5] Barker EB, Chen L, Roginsky AL, Vassilev A, Davis R, Simon S (2019) Recommendation for Pair-Wise Key-Establishment Using Integer Factorization Cryptography. (National Institute of Standards and Technology, Gaithersburg, MD), NIST Special Publication (SP) 800-56B, Rev. 2. https://doi.org/10.6028/NIST.SP.800-56Br2.

[6] National Institute of Standards and Technology (2015) SHA-3 Standard: Permutation-Based Hash and Extendable-Output Functions. (Department of Commerce, Washington, DC), Federal Information Processing Standards Publication (FIPS) NIST FIPS 202. https://doi.org/10.6028/NIST.FIPS.202.

[7] Kelsey JM, Chang SjH, Perlner RA (2016) SHA-3 Derived Functions: cSHAKE, KMAC, TupleHash and ParallelHash. (National Institute of Standards and Technology, Gaithersburg, MD), NIST Special Publication (SP) 800-185. https://doi.org/10.6028/NIST.SP.800-185.

[8] National Institute of Standards and Technology (2015) Secure Hash Standard (SHS). (Department of Commerce, Washington, D.C.), Federal Information Processing Standards Publication (FIPS) NIST FIPS 180-4. https://doi.org/10.6028/NIST.FIPS.180-4.

[9] National Institute of Standards and Technology (2024) Recommendation for Key Management: Part 1 – General. (National Institute of Standards and Technology, Gaithersburg, MD), NIST Special Publication (SP) 800-57 Part 1, Rev 6. [Forthcoming: will be available at https://csrc.nist.gov/publications].

[10] Aumasson JP, Bernstein DJ, Beullens W, Dobraunig C, Eichlseder M, Fluhrer S, Gazdag SL, Hülsing A, Kampanakis P, Kölbl S, Lange T, Lauridsen MM, Mendel F, Niederhagen R, Rechberger C, Rijneveld J, Schwabe P, Westerbaan B (2022) SPHINCS$^+$ – Submission to the NIST post-quantum project, v.3.1. Available at https://sphincs.org/data/sphincs+-r3.1-specification.pdf.

[11] Hülsing A, Butin D, Gazdag SL, Rijneveld J, Mohaisen A (2018) XMSS: eXtended Merkle Signature Scheme. (Internet Research Task Force (IRTF)), IRTF Request for Comments (RFC) 8391. https://doi.org/10.17487/RFC8391.

[12] Cooper DA, Apon D, Dang QH, Davidson MS, Dworkin MJ, Miller CA (2020) Recommendation for Stateful Hash-Based Signature Schemes. (National Institute of Standards and Technology, Gaithersburg, MD), NIST Special Publication (SP) 800-208. https://doi.org/10.6028/NIST.SP.800-208.

[13] Merkle RC (1979) *Secrecy, Authentication, and Public Key Systems*. Ph.D. thesis. Stanford university, .

[14] Barker EB, Kelsey JM (2015) Recommendation for Random Number Generation Using Deterministic Random Bit Generators. (National Institute of Standards and Technology, Gaithersburg, MD), NIST Special Publication (SP) 800-90A, Rev. 1. https://doi.org/10.6028/NIST.SP.800-90Ar1.

[15] Sönmez Turan M, Barker EB, Kelsey JM, McKay KA, Baish ML, Boyle M (2018) Recommendation for the Entropy Sources Used for Random Bit Generation. (National Institute of Standards and Technology, Gaithersburg, MD), NIST Special Publication (SP) 800-90B. https://doi.org/10.6028/NIST.SP.800-90B.

[16] Barker EB, Kelsey JM, McKay KA, Roginsky AL, Sönmez Turan M (2022) Recommendation for Random Bit Generator (RBG) Constructions. (National Institute of Standards and Technology, Gaithersburg, MD), NIST Special Publication (SP) 800-90C 4pd. https://doi.org/10.6028/NIST.SP.800-90C.4pd.

[17] Kannwischer MJ, Genêt A, Butin D, Krämer J, Buchmann J (2018) Differential Power Analysis of XMSS and SPHINCS. *Constructive Side-Channel Analysis and Secure Design*, eds Fan J, Gierlichs B (Springer International Publishing, Cham), pp 168–188. https://doi.org/10.1007/978-3-319-89641-0_10.

[18] Castelnovi L, Martinelli A, Prest T (2018) Grafting Trees: A Fault Attack Against the SPHINCS Framework. *Post-Quantum Cryptography*, eds Lange T, Steinwandt R (Springer International Publishing, Cham), pp 165–184. https://doi.org/10.1007/978-3-319-79063-3_8.

[19] Genêt A, Kannwischer MJ, Pelletier H, McLauchlan A (2018) Practical Fault Injection Attacks on SPHINCS, *Cryptology ePrint Archive preprint*. https://ia.cr/2018/674.

[20] Amiet D, Leuenberger L, Curiger A, Zbinden P (2020) FPGA-based SPHINCS+ Implementations: Mind the Glitch. *2020 23rd Euromicro Conference on Digital System Design (DSD)*, pp 229–237. https://doi.org/10.1109/DSD51259.2020.00046.

[21] Genêt A (2023) On Protecting SPHINCS+ Against Fault Attacks. *IACR Transactions on Cryptographic Hardware and Embedded Systems* 2023(2):80–114. https://doi.org/10.46586/tches.v2023.i2.80-114.

[22] Groot Bruinderink L, Hülsing A (2018) "Oops, I Did It Again" – Security of One-Time Signatures Under Two-Message Attacks. *Selected Areas in Cryptography – SAC 2017*, eds Adams C, Camenisch J (Springer International Publishing, Cham), pp 299–322. https://doi.org/10.1007/978-3-319-72565-9_15.

48

[23] Housley R (2009) Cryptographic Message Syntax (CMS). (Internet Engineering Task Force (IETF)), IETF Request for Comments (RFC) 5652. https://doi.org/10.17487/RFC5652.

[24] National Institute of Standards and Technology (2024) Recommendation for Additional Stateless Hash-Based Digital Signature Parameter Sets. (National Institute of Standards and Technology, Gaithersburg, MD), NIST Special Publication (SP) 800-230. [Forthcoming: will be available at https://csrc.nist.gov/publications].

[25] National Institute of Standards and Technology (2016) Submission Requirements and Evaluation Criteria for the Post-Quantum Cryptography Standardization Process. Available at https://csrc.nist.gov/CSRC/media/Projects/Post-Quantum-Cryptography/documents/call-for-proposals-final-dec-2016.pdf.

[26] Alagic G, Apon D, Cooper DA, Dang QH, Dang T, Kelsey JM, Lichtinger J, Liu YK, Miller CA, Moody D, Peralta R, Perlner RA, Robinson A, Smith-Tone D (2022) Status Report on the Third Round of the NIST Post-Quantum Cryptography Standardization Process. (National Institute of Standards and Technology, Gaithersburg, MD), NIST Interagency or Internal Report (IR) NIST IR 8413-upd1, includes updates as of September 26, 2022. https://doi.org/10.6028/NIST.IR.8413-upd1.

[27] Hülsing A, Kudinov M (2022) Recovering the Tight Security Proof of SPHINCS$^+$. *Advances in Cryptology – ASIACRYPT 2022*, eds Agrawal S, Lin D (Springer Nature Switzerland, Cham), pp 3–33. https://doi.org/10.1007/978-3-031-22972-5_1.

[28] Stern J, Pointcheval D, Malone-Lee J, Smart NP (2002) Flaws in Applying Proof Methodologies to Signature Schemes. *Advances in Cryptology — CRYPTO 2002*, ed Yung M (Springer Berlin Heidelberg, Berlin, Heidelberg), pp 93–110. https://doi.org/10.1007/3-540-45708-9_7.

[29] Cremers C, Düzlü S, Fiedler R, Janson C, Fischlin M (2021) BUFFing Signature Schemes Beyond Unforgeability and the Case of Post-Quantum Signatures. *2021 IEEE Symposium on Security and Privacy (SP)* (IEEE Computer Society, Los Alamitos, CA, USA), pp 1696–1714. https://doi.org/10.1109/SP40001.2021.00093.

[30] Moriarty K, Kaliski B, Jonsson J, Rusch A (2016) PKCS #1: RSA Cryptography Specifications Version 2.2. (Internet Engineering Task Force (IETF)), IETF request for comments (RFC) 8017. https://doi.org/10.17487/RFC8017.

[31] National Institute of Standards and Technology (2008) The Keyed-Hash Message Authentication Code (HMAC). (Department of Commerce, Washington, DC), Federal Information Processing Standards Publication (FIPS) NIST FIPS 198-1. https://doi.org/10.6028/NIST.FIPS.198-1.

[32] Krawczyk H, Bellare M, Canetti R (1997) HMAC: Keyed-Hashing for Message Authentication. (Internet Engineering Task Force (IETF)), IETF request for comments (RFC) 2104. https://doi.org/10.17487/RFC2104.

[33] Stern M (2021) Re: Diversity of signature schemes. Available at https://groups.google.com/a/list.nist.gov/g/pqc-forum/c/2LEoSpskELs/m/LkUdQ5mKAwAJ.

[34] Antonov S (2022) ROUND 3 OFFICIAL COMMENT: SPHINCS+. Available at https://groups.google.com/a/list.nist.gov/g/pqc-forum/c/FVItvyRea28/m/mGaRi5iZBwAJ.

[35] Perlner R, Kelsey J, Cooper D (2022) Breaking Category Five SPHINCS$^+$ with SHA-256. *Post-Quantum Cryptography*, eds Cheon JH, Johansson T (Springer International Publishing, Cham), pp 501–522. https://doi.org/10.1007/978-3-031-17234-2_23.

Appendix A — Differences From the SPHINCS$^+$ Submission

This standard is based on Version 3.1 of the SPHINCS$^+$ specification [10] and contains several minor modifications compared to Version 3 [4], which was submitted at the beginning of round three of the NIST PQC Standardization process:

- Two new address types — WOTS_PRF and FORS_PRF — were defined for WOTS$^+$ and FORS secret key value generation.

- $\mathbf{PK}$.seed was added as an input to $\mathbf{PRF}$ in order to mitigate multi-key attacks.

- For the category 3 and 5 SHA2 parameter sets, SHA-256 was replaced by SHA-512 in $\mathbf{H}_{msg}$, $\mathbf{PRF}_{msg}$, $\mathbf{H}$, and $\mathbf{T}_\ell$ based on weaknesses that were discovered when using SHA-256 to obtain category 5 security [33, 34, 35].

- R and $\mathbf{PK}$.seed were added as inputs to MGF1 when computing $\mathbf{H}_{msg}$ for the SHA2 parameter sets in order to mitigate multi-target long-message second preimage attacks.

This standard also differs from the Version 3 specification in its method for extracting bits from the message digest to select a FORS key. This change was made in order to align with the reference implementation that was submitted along with the round three specification. The description of the method for extracting indices for FORS signature generation and verification from the message digest was also changed due to ambiguity in the submitted specification. The method described in this standard is not compatible with the method used in the reference implementation that was submitted along with the round three specification. Additionally, line 6 in both wots_sign and wots_pkFromSig were changed to match the reference implementation, as the pseudocode in [10, 4] will sometimes shift $csum$ by the incorrect amount when lg_w is not 4.

This standard **approves** the use of only 12 of the 36 parameter sets defined in [10, 4]. As specified in Section 11, only the 'simple' instances of the SHA2 and SHAKE parameter sets are **approved**.

A.1   Changes From FIPS 205 Initial Public Draft

The differences from Version 3 of the SPHINCS$^+$ specification described in Appendix A were included in the draft version of this standard (FIPS 205 ipd) that was posted on August 24, 2023. Based on comments that were submitted on FIPS 205 ipd, the SLH-DSA signature generation and verification functions were modified to include domain separation cases in which the message is signed directly and in which a digest of the message is signed. The changes were made by modifying the inputs to the signing and verification functions (see Algorithms 22, 23, 24, and 25).